Groundwater Base Level Changes and Adjoining Hydrological Systems

Uri Kafri • Yoseph Yechieli

Groundwater Base Level Changes and Adjoining Hydrological Systems

Springer

Dr. Uri Kafri
Geological Survey of Israel
95 501 Jerusalem
Israel
uri.kafri@gsi.gov.il

Dr. Yoseph Yechieli
Geological Survey of Israel
95 501 Jerusalem
Israel
yechieli@gsi.goc.il

ISBN 978-3-642-13943-7 e-ISBN 978-3-642-13944-4
DOI 10.1007/978-3-642-13944-4
Springer Heidelberg Dordrecht London New York

Library of Congress Control Number: 2010934210

Cover illustration: An enlarged sattelite photo of the endorheic Dead Sea Rift base level and a sattelite photo of the Mediterranean Sea base level and the nearby Dead Sea Rift base level (lower right corner)

Cover design: deblik, Berlin

Printed on acid-free paper

Springer is part of Springer Science+Business Media (www.springer.com)

Acknowledgements

The Geological Society of America is acknowledged for granting permission to use Figs. 3.1, 4.2, 4.5, 5.2, 7.1, 9.4, 9.10, 9.11, 10.4, 10.6, 10.7, 11.3, 11.25.

Wiley-Blackwell is acknowledged for granting permission to use Figs. 3.3, 3.4, 4.1, 5.1, 7.5, 9.7.

Sage Publications is acknowledged for granting permission to use Fig. 9.2.

The American Association of Petroleum Geologists is acknowledged for granting permission to use Fig. 11.2.

Thanks are due to Nili Almog and Bat-Sheva Cohen of the Graphics and Publishing Drafting Division of the Geological Survey of Israel for their assistance in the graphics.

Contents

Chapter 1
Introduction

The main objective of this book is to trigger a fact finding effort and an interdisciplinary approach to the role of groundwater base-levels in hydrogeological systems due to their different configurations and their changes in time and space.

The term "base-level" is usually related to surface water drainage systems and erosion processes. Several definitions exist to this term in the literature, one of which is "The lowest level to which a land surface can be reduced by the action of running water". In general, the groundwater base-level was described as "a drainage level for the aquifer that represents the lowest groundwater level that will occur from groundwater flow only" (Olin 1995). It is herein, in analogy, defined as the ultimate discharge zone down-gradient of groundwater basin. Groundwater base levels are basically grouped in two types: (a) the global marine base-level, and (b) the continental, terminal, internal drainage system, closed basin or termed as an endorheic base-level that attract and discharge convergent groundwater flow. An endorheic base level is usually described as a closed drainage basin that retains water and allows no outflow to other bodies of water. Precipitation that falls within the basin does not flow out and may only leave the drainage system by evaporation or seepage.

Groundwater basins and surface water basins that drain, approximately, the same region do not necessarily coincide. In some cases they can share roughly the same boundaries, divides and base-levels, whereas in other cases the same area can be drained by surface networks to a specific base-level and by groundwater systems to a different one. Endorheic surface water continental base-levels are relatively easily defined and delineated based on their adjoining surface drainage network system. Some endorheic groundwater base-levels, on the other hand, are not always easily delineated due to the scarcity of groundwater table data. Still, the interconnection between terminal surface water base-levels and lakes and adjoining groundwater systems were analyzed and established among others by Smith et al. (1997), Urbano et al. (2000, 2004), Holzbecher (2001), Gosselin and Khisty (2001), Winter (1999), Winter et al. (2002) and Pederson (2001). It is noticed that levels of many lakes that serve as surface water base-levels are also controlled by groundwater levels and the lakes serve also as groundwater base-levels (see also Sect. 5.2)

U. Kafri and Y. Yechieli, *Groundwater Base Level Changes and Adjoining Hydrological Systems*,
DOI 10.1007/978-3-642-13944-4_1, © Springer-Verlag Berlin Heidelberg 2010

An attempt is made herein to summarize, compile and compare results of studies, sometimes partial, regarding current groundwater base-levels as well as of paleo-ones, still within the hydrological "memory" time window, mostly from Neogene times to the present.

The book aims to propose a basic classification of groundwater base-levels, both marine and continental. The latter includes terminal and flow-through (intermediate) base-levels. Special attention is given to continental endorheic base-levels considerably below sea level whether connected hydrologically to the sea or not.

Base levels are subjected to changes in time and space. The factors that control base-level changes are also discussed in the book. Among these are: tectonic movements, isostatic changes as well as glacio-isostatic rebound of continents following deglaciation, eustatic sea level and lake level changes caused by climate changes, water table depressions caused by evapotranspiration, compaction, and capturing of groundwater basins. Some emphasis is also given to the impact of the claimed on-going global warming process, whether natural or anthropogenic (the green house effect) on base level changes. Changes are also caused by land subsidence due to man-made groundwater withdrawal or hydrocarbon exploitation.

The different methods and devices employed for measuring and monitoring current base-level changes are also described. Among these are marine gauges and mariographs to monitor sea and lake level changes, geodetic levelings by the GPS and INSAR methods to detect ground level changes, and monitoring devices to measure the response of the groundwater system. The different methods to reconstruct paleo and ancient base-level changes are also described. Among these are shallow and near shore marine and lake sediments, abrasional notches and platforms, paleo-karstic systems, location of archaeological sites, as well as the use of both radiogenic and stable isotopes and hydrological modeling.

Among other important topics that are discussed are Neogene to Holocene paleo- base- levels and their adjoining hydrological systems. In addition, the impact of base-level changes on capturing of groundwater basins and shift of divides are also discussed.

The configuration and changes of groundwater base-levels play a most important role in the response of the groundwater systems regarding salination and flushing processes. The discussed topics, concerning this matter, are: salination and flushing of coastal aquifers, salination of terminal continental base-levels, below sea level and connected hydraulically to the sea as well as density driven salination processes that occur in terminal continental base-levels, with no connection to the sea.

All the topics described above are accompanied and supported by selected examples and references from all over the world. Special emphasis is paid in this book to examples from the Mediterranean marine base-level and from the endorheic Dead Sea Rift base-level of Israel. This is due to the relatively high abundance and availability of data, as well as to the long term involvement of the authors in the study of these base-levels, that can serve as analogues to other base-levels where data do not exist or are scarce.

The present book does not pretend to list and describe all the existing and paleo base levels on earth. It attempts to describe the most known ones and particularly those that render abundant available hydrogeological data. In some of the described cases, interpretation is made as to the nature of the described base levels, based primarily on the resemblance to other ones, despite their scarce available data.

References

Gosselin DC, Khisty MJ (2001) Simulating the influence of two shallow, flow-through lakes on a groundwater system: implications for groundwater mounds and hinge lines. Hydrogeol J 9:476–486

Holzbecher E (2001) The dynamics of subsurface water divides- watersheds of Lake Stechlin and neighbouring lakes. Hydrol Process 15:2297–2304

Olin M (1995) Estimation of base level for an aquifer from recession rates of groundwater levels. Hydrogeol J 3:40–51

Pederson DT (2001) Stream piracy revisited: a groundwater sapping solution. GSA Today 11:4–10

Smith AJ, Donovan JJ, Ito E, Engstrom DR (1997) Ground-water processes controlling a prairie lake's response to middle Holocene drought. Geology 25:391–394

Urbano LD, Person M, Hanor J (2000) Groundwater–lake interactions in semi-arid environments. J Chem Explor 69–70:423–427

Urbano LD, Person M, Kelts K, Hanor JS (2004) Transient groundwater impacts on the development of paleoclimatic lake records in semi-arid environments. Geofluids 4:187–196

Winter TC (1999) Relation of streams, lakes, and wetlands to groundwater flow systems. Hydrogeol J 7:28–45

Winter TC, Harvey, JW, Franke OL, Alley WM (2002) Groundwater and surface water – a single resource. US Geol Surv circular 1139:1–79

Chapter 2
The Main Types of Groundwater Base-Level

There are basically two main types of groundwater base-levels, namely the marine ones and the continental ones.

The marine base-levels include the oceans and their connected seas and in places coastal sabkhas. They serve as terminal base-level to groundwater flow, as well as to surface flow from the continents. Being interconnected, and different from continental base-level, they have roughly the same level which is the global sea level. Climate changes, in the past, have affected globally the sea levels.

Continental base-levels are geographically defined and described from all over the world. They consist of terminal, termed endorheic, closed systems that are not interconnected in most cases. They differ, regarding their dimensions, between large, regional, base-levels and smaller local ones. Regarding their elevation, some of them are above sea level, whereas some are below sea level and even considerably below it. The latter category includes two types, that assumingly have a different relevance regarding salination processes, namely those base-levels that are distant from (Chap. 10) or those close to, and possibly connected to the sea (Chap. 11).

Local continental base-levels are also formed naturally where local sabkhas or areas that occupy phreatophytes, occur as discharge zones, resulting in convergent groundwater flow toward them and a water table sink due to high evapotranspiration (Sect. 3.6). Local man-made hydrological sinks are also formed through groundwater dewatering or over-exploitation or hydrocarbon abstraction (Sect. 3.7).

A special category are the flow-through intermediate base-levels that are not terminal or endorheic by definition. These base-levels attract convergent groundwater flow but they still drain constantly or temporarily by surface and/or subsurface flows down stream to a down gradient marine base-level or to a terminal continental one.

The above listed base-level types, as schematically shown on Fig. 2.1, occur at present and occurred also in the past as different paleo-base- level types.

U. Kafri and Y. Yechieli, *Groundwater Base Level Changes and Adjoining Hydrological Systems*,
DOI 10.1007/978-3-642-13944-4_2, © Springer-Verlag Berlin Heidelberg 2010

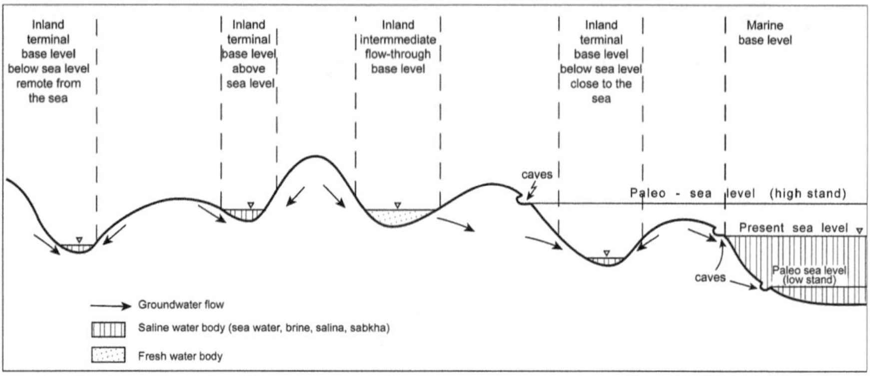

Fig. 2.1 Schematic description of main base level types

The different base-level types are analyzed further in this book, accompanied by detailed, as available, description of examples from all over the world (see also Chaps. 7–9).

Chapter 3
Factors Controlling Base-Level Elevation Changes

3.1 General

Groundwater base-level changes can stem from the drop or rise of the base-level or the subsidence or uplift of the adjoining groundwater basin. In some cases, a combination of both is also possible. The main factors that control variations in groundwater base-level elevation are detailed below.

3.2 Global, Regional and Local Climate Changes

Global climate changes have been known to occur throughout the geological history, at least since the last 500 Ma, an episode known as the "snowball effect" (Hoffman et al. 1998). More detailed records of climatic changes are available for the period since the end of the Cretaceous, some 66 Ma, when several extreme climate change events caused the extinction of many species. The best continuous record for climatic changes is known for the Pleistocene Era. These changes are attributed to changes in solar radiation (Milankovitch cycles) and atmospheric circulation, resulting in cold and warmer periods or glacial and inter-glacial ones respectively. During glacial periods huge volumes of ice were accumulated on the poles as well as on the high altitudes in the continents on the expense of the ocean waters and as a result the oceanic and connected seas (base) levels declined. During subsequent warmer, inter-glacial periods the snow and ice had melt adding huge volumes of water to the oceans, thereby resulting in a rise of the oceanic and the connected marine base-levels. These changes, termed as eustatic changes, were discussed and described for the different geological periods and from various places in the world, among others, by Haq et al. (1987) and Sahagian and Holland (1991). Kominz et al. (1998), for example, show a long term sea-level curve of the Cenozoic Era exhibiting considerable long term sea-level fluctuations (Fig. 3.1).

U. Kafri and Y. Yechieli, *Groundwater Base Level Changes and Adjoining Hydrological Systems*,
DOI 10.1007/978-3-642-13944-4_3, © Springer-Verlag Berlin Heidelberg 2010

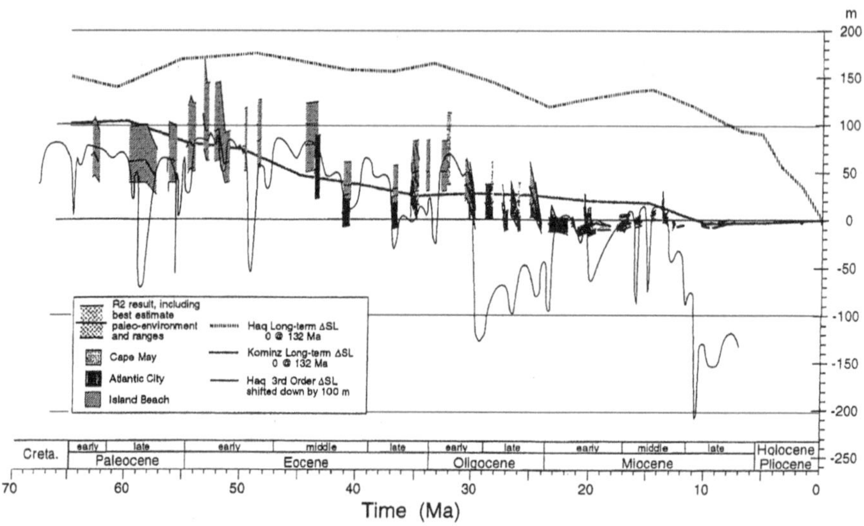

Fig. 3.1 Long term global sea level curves [after Kominz et al. (1998)]

The considerable drop of the eustatic sea-level in the Last Glacial Maximum (LGM) to approximately 120 m below sea level (bsl), and its subsequent rise to the present day sea-level since then, was discussed in numerous studies. A global eustatic sea level curve was exhibited by Fleming et al. (1998) for the last part of the Pleistocene (since 200 ka) and for the period since 140 ka (Fig. 3.2). Fairbanks (1989) also described the eustatic sea level changes for the last 17 ka. It is interesting to note that, due to the current climate change and the described on-going global warming process which is now well established (e.g., IPCC 2007), a global sea level rise is expected in the future. It was claimed that global warming and sea level rise will probably continue even if the concentration of the green-house gases, which contribute to the warming, is stabilized (Meehl et al. 2005). In the eastern Mediterranean, for example, sea level rose by 0.5–1.0 cm/y during 1993–2005 and a further sea level rise of ~0.5 m is expected by the end of the twenty-first century (Warrick et al. 1996). Such a sea level rise will assumingly affect the hydrology of the low lying Pacific or Maldivian islands and atolls as well as flat deltas such as the Nile and the Bengal Bay deltas (Sherif and Singh 1999). It should be mentioned that some studies (i.e., Chilingar et al. 2009) claim that the rising concentration of greenhouse gases may result in the cooling of climate and that the anthropogenic contribution of these gases has no practical effect on climate.

Inland seas and lakes serve as flow-through, intermediate, or endorheic terminal base-levels to both surface and groundwater flows. Their levels are sensitive to global climatic changes, from dry to humid, during respective glacial and interglacial periods caused by changes of global atmospheric circulation. Inland seas and lakes can be sensitive also to local climate changes. The climatic changes result in expansion or shrinkage of paleo water bodies and changes in their levels

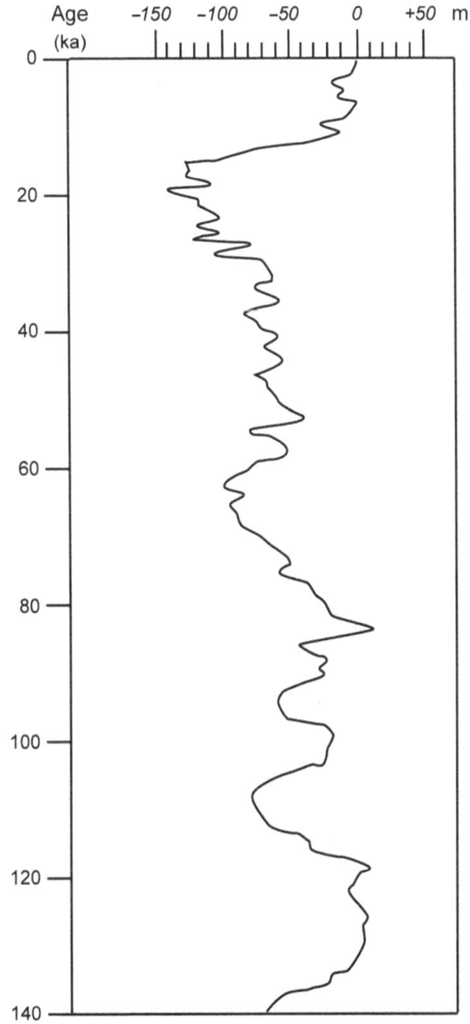

Fig. 3.2 Global sea levels since 140 ka [modified after Waldmann et al. (2009)]

as manifested by multiple systems of paleo shore lines. Examples to the above changes were described from all over the world, and are discussed further in this book, as follows: From north Africa and the Mediterranean (Griffin 2002); from lakes in central Asia and China (Qin and Yu 1998; Yang et al. 2004) (see also Sect. 10.4); from Siberia (Borodavko et al. 2003); from the Caspian Sea (Rychagov 1997) (see also Sect. 10.5); from African lakes (Scholz and Rosendahl 1988; Trauth et al. 2003; Garcin et al. 2008); from Lake Chad (Helfert and Holz 1985; Drake and Bristow 2006; Leblanc et al. 2007) (see also Sect. 9.2); from the Death Valley (Jianren et al. 1996; Lowenstein et al. 1999; Miner et al. 2007) (see also Sect. 10.3); from Mexico (Castiglia and Fawcett 2006); from Lake Eyre (DeVogel et al. 2004; Magee et al. 2004) (see also Sect. 10.2) and from the Dead Sea Basin (Stein 2001) (see also Sect. 11.2). Due to the above changes, these paleo water bodies were also subjected to changes in their salinity as well as their chemical and isotopic composition, as recorded in their deposited sediments (i.e., Jianren et al. 1996).

Such changes, associated also with tectonic movements, could be in some cases, more than just a minor rise or fall of the water level. The Paratethys and Mediterranean basins, for example, were subjected to major base-levels changes in Miocene to Pliocene times, being diverted from a base-level connected to the global marine system to closed inland seas as described in detail, for example, by Popov et al. (2006) (see also Sect. 7.2). Similarly, the Black Sea disconnection from, or connection to the global marine system, via the Mediterranean Sea, were controlled by the drastic global sea level changes since the LGM (see also Sect. 7.3).

3.3 Tectonic Movements

Tectonic movements are one of the main factors controlling both marine and continental base-level changes. Tectonic movements of the earth crust are known to raise or lower the ocean basins and deform the continents. Broad epirogenic movements usually uplift the continents and press the floor of the ocean basins, thereby lowering the sea (base) level.

Orogenic movements are associated with mountain building processes at the convergence of tectonic plates. This phenomenon is abundant in ocean basins and seas, characterized by uplift, subsidence and deformation of the ocean basin margins, thus changing base-level elevations. Examples to the above process are: the regional uplift connected to the subduction zone in the Pacific coast of Mexico (Ramirez-Herrera and Urrutia-Fucugauchi 1999); the uplift of Patagonia, Argentina (Rostami et al. 2000); the closure of the Paratethys and the closure of the Mediterranean Sea in the Neogene followed by the drop of its level (Popov et al. 2006) (see also Sect. 7.2).

Regional or local tectonic movements of uplift, subsidence or rifting along shorelines or on-land are also responsible for level changes of continental, both lacustrine and non-aquatic depressions that act as base-levels. Examples are: detachment from the Mediterranean, closure and diversion to a Pliocene inland receding lake in the Dead Sea Rift (Kafri and Ecker 1964) (see also Sect. 11.2); tectonic uplift of shorelines in New Zealand (Williams 1982), in Corinth, Greece (Collier 1990), in Calabria, Italy (Cucci 2004; Dumas et al. 2006) and Lake Baikal (Colman 1998; Kolomiets 2008).

3.4 Isostatic and Glacio-Isostatic Movements

Isostatic movements take place as an adjustment process of the earth crust as a result of loading and unloading of the surface. Accumulation of sediments or volcanic deposits in certain areas such as sea margins, deltas or continental depressions often results in subsidence because of the loading. The same process occurs during transgressions through the deepening of continental shelves. The increasing hydrostatic pressure of the water column depresses the continental shelf, thereby effecting the base-level elevation.

Another known isostatic phenomenon is the glacio-isostatic process.

Glacio-isostatic processes play an important role in the northern hemisphere on oceanic base-level changes relative to their adjoining on-land groundwater systems.

The well known and well described example is the glacio-isostatic rebound of the Fennoscandian Shield since the beginning of the deglaciation following the peak of the LGM some 20 ka BP (i.e., Fjeldskaar et al. 2000). This process resulted in a continuous, still ongoing, major uplift above sea level and subsidence below sea level in the margins of the uplifted continental area. The process was affected by both the unloading of the wide-spread ice sheets of northern Europe and the stress field of the plates motion. Both resulted in additional neotectonic movements and seismic activity (Gregersen 2006).

The post-glacial uplift, evidenced mainly by ancient shore lines, started around 13 ka BP, amounting to a total of 830 m in the center of Fennoscandia (Fig. 3.3) (Moerner 1991). The long-term post-glacial rate of uplift was found out to be close to 5 mm/y (Moerner 1991). The interplay between the uplift of the land and the sea level rise through the snow melt is also well manifested in the British Isles. The first

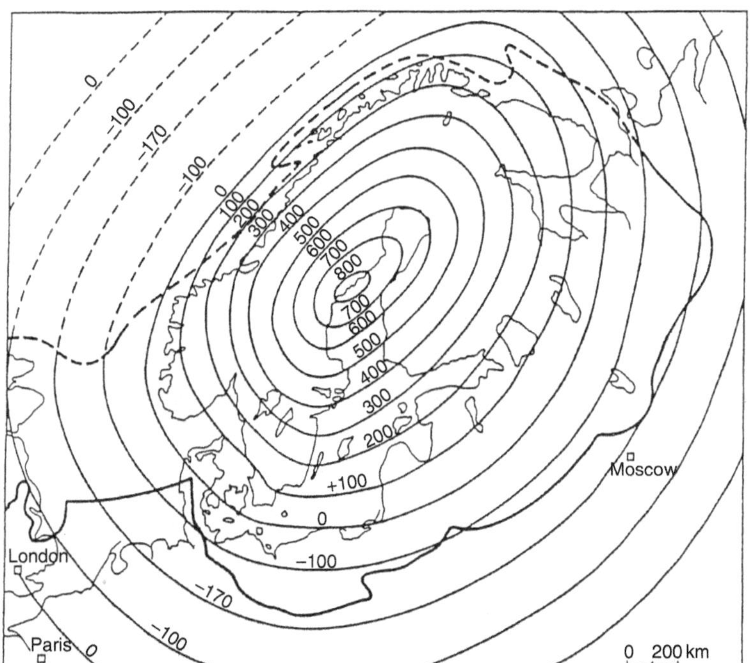

Fig. 3.3 Total absolute uplift (in meters) of the Fennoscandian Shield since the LGM [after Moerner (1991)]

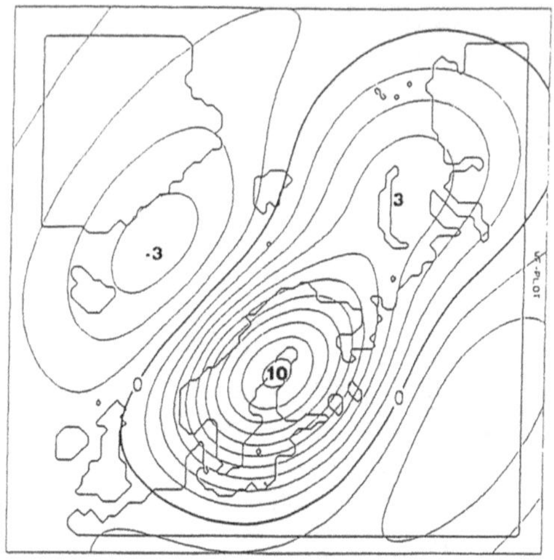

Fig. 3.4 Theoretical present rate (in millimeter/year) of uplift of Fennoscandia [after Moerner (1991)]

is more effective as you get closer to the center of uplift whereas the latter becomes more effective with distance from the center (Lambeck 1991).

The current rate of uplift of Fennoscandia is obtained by calculating repeated measurements of geodetic precise levelings, GPS and VLBI interferometry, tide-guages and gravimetry (see also Sect. 4.1). The present rate of uplift in the center of uplift attains a value of 8 mm/y (Fjeldskaar et al. 2000). Theoretical calculations, based on different mantle and asthenosphere viscosities yielded similar uplift rates, between 10 and 14 mm/y (Fig. 3.4), (Fjeldskaar and Cathles 1991). Similar to the Fennoscandian case, Baedke and Thompson (2000) also described the changes of the old Lake Michigan levels as a result of the glacio-isostatic process.

3.5 Land Subsidence Through Compaction and Dewatering

Land subsidence in sedimentary basins may occur as a result of both natural and anthropogenic (man-made) induced processes of compaction and settling of the sediments involved.

Natural compaction of argillaceous sediments was studied and discussed, among numerous researchers, by Rieke and Chilingar (1974). Natural compaction of basin sediments in response to sediment loading is responsible for land subsidence. As an example, a case from the coastal area of the Netherlands was described and modeled, yielding rates of subsidence between 0.1 and 1 mm/y (Kooi and de Vries 1998). The same process in the Nile Delta yielded subsidence rates as much as 0.5 mm/y (Kooi and Groen 2003). Consolidation of peat sediments was found to be responsible for land subsidence in the Potina coastal plain of Italy (Brunamonte et al. 2000).

Land subsidence caused by sediment loading, lignite maturation and water loss through compaction and consolidation was described from the Hula Basin, Israel (Kafri et al. 1983). Baer et al. (2002) described the process of land subsidence in the coastal plain of the Dead Sea, Israel, through consolidation following the dewatering of the coastal aquifer related to the continuous current drop of the Dead Sea level (see also Sect. 11.2.2).

Land subsidence is also caused by compaction due to man-made groundwater and hydrocarbon extraction. The process, related to groundwater level drop, was described, among others, by Terzaghi (1925). The drop of the water level results in a decrease of pore pressure in the aquifer system. Consequently, the effective stress (overburden minus pore pressure) increases and the pores structure in the fine grained material is re-arranged so as to reduce pore space, which might, in turn, cause consolidation, collapse and a resultant land subsidence.

Modeling and prediction of land subsidence was carried out, among others, by Monjoie et al. (1992) and Liu and Helm (2008). Land subsidence through groundwater extraction was studied and described from the United States (Bull 1973; Galloway et al. 1998; Sun et al. 1999; Carruth et al. 2007), from China (Shearer 1998; Ma et al. 2006; Zhang et al. 2007; Shi et al. 2008) and from Italy (Rossi et al. 2000; Teatini et al. 2006). Land subsidence through deep fluid extraction was also described from the Cerro Prieto pull apart basin and geothermal field, located in the Mexicali Valley, within the Salton Trough (Glowasca et al. 2010) (see also Sect. 11.5).

Land subsidence through hydrocarbon production was described, for example, from Italy (Bertoni et al. 2000; Bau et al. 2001), from the Netherlands (Houtenbos 2000) and from the United States (Chan and Zoback 2007).

The combined effect of natural compaction and water and gas withdrawal on land subsidence was described from the northern Adriatic coastland of Italy by Gambolati and Teatini (2000).

3.6 Groundwater Cones of Depression Caused by Natural Processes of Evapotranspiration

Evapotranspiration through vegetation is a major withdrawal factor in groundwater balance. In the case of arid or semi-arid environments, coupled with very shallow water tables, the role of phreatophytes might become regionally or locally very important. Phreatophytes are plants whose roots reach into the capillary fringe above the water table. Small cones of depression in the water table around phreatophytes are sometimes formed due to the removal of groundwater by the roots (Fig. 3.5) (Winter et al. 2002). A similar phenomenon of convergent upward shallow groundwater flow to wetland areas can also create local cones of depression due to diurnal or seasonal fluctuations in the water levels of lakes or wetlands as a result of evapotranspiration (Fig. 3.6) (i.e., Winter 1999; Schilling 2007).

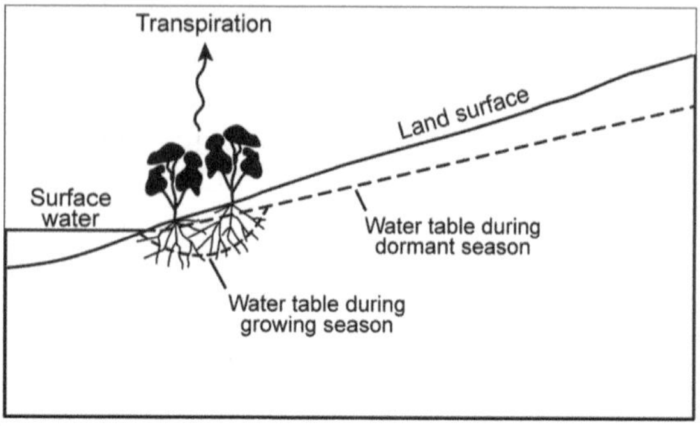

Fig. 3.5 The formation of a water table cone of depression by phreatophytes [after Winter et al. (2002)]

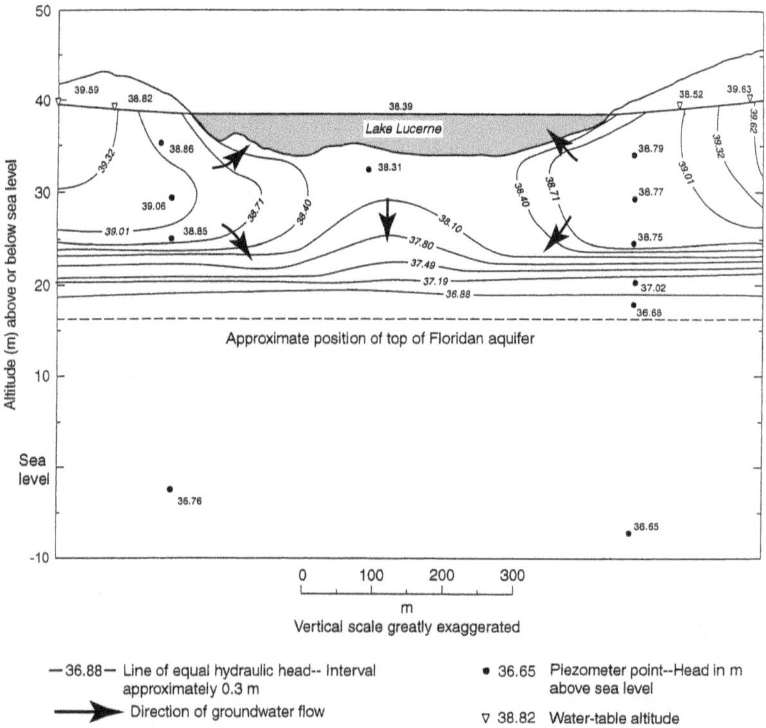

Fig. 3.6 Hydrologic section exhibiting convergent groundwater flow to a lake [after Winter (1999)]

Leblanc et al. (2007) described, from the Quaternary aquifer of the Lake Chad basin, large piezometric depressions or vast concentric sinks also called "hollow aquifers". These depressions have an amplitude of about 40 m and are found in vast areas around Lake Chad. This phenomenon was related by them to evapotranspiration (see also Sect. 9.2).

Depending on the dimension of the above phenomena, such cones of depression might serve as local base-levels or discharge zones resulting in diversion or capturing of groundwater flow similar to the groundwater extraction case (see also Sect. 3.7).

3.7 Groundwater Cones of Depression Caused by Groundwater Exploitation

Over-pumping or over-exploitation of groundwater on a regional or local scale often creates a cone of depression in the water table sometimes to considerable depths. Such depressions can attain a depth of several tens of meter and even a 100 m (i.e., Dintelmann 2004). These cones of depression, whether of local extent or a regional one, act as on-land artificial discharge zones through pumping and a base-level that attracts convergent groundwater flows. The dimensions of such cones change as a function of the amount of exploitation or the recovery of the depleted water table. The dimensions also depend on whether the exploited aquifer is phreatic or confined. These processes involve also capturing and diversion of groundwater flow.

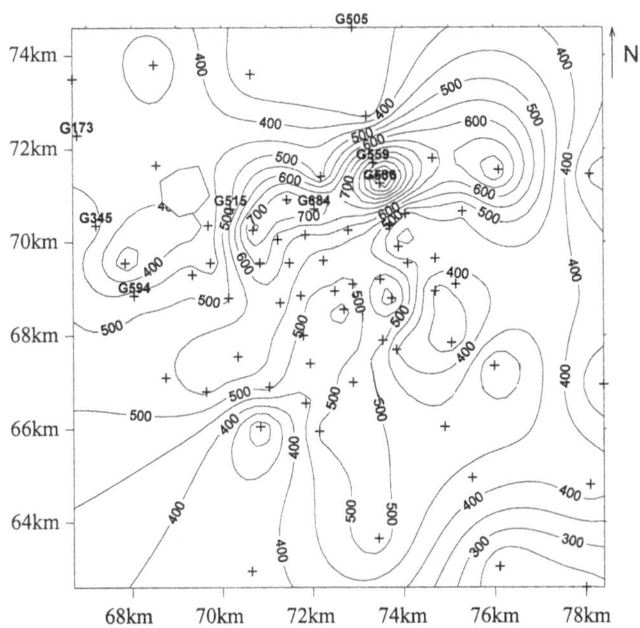

Fig. 3.7 The amount of land subsidence (in millimeter, based on survey in 1983–1997) in the Suzhou City, China due to dewatering [after Chen et al. (2003)]. The *crosses* are the location of the survey points

The excessive groundwater exploitation and the resultant water table draw-down causes, in turn, land subsidence phenomena in many places all over the world as reported before, for example, from California (Larson et al. 2001) or from China (Fig. 3.7) (Chen et al. 2003; Shi et al. 2008).

When over-exploitation occurs in coastal aquifers the outcome is often the creation of water table cones of depression close to the sea and the water table drops to considerably below sea level. This, in turn, might enhance sea water encroachment into the aquifers and their salination by upconning of seawater into the system. Numerous cases of the above were described from all over the world, such as in Mexico (Steinich et al. 1998), China (Liu et al. 2001; Zhou et al. 2007), Israel (Schmorak and Mercado 1969) and Greece (Petalas and Lambrakis 2006).

References

Baedke SJ, Thompson TA (2000) A 4,700-year record of lake level and isostasy for lake Michigan. J Great Lakes Res 26:416–426

Baer G, Schattner U, Wachs D, Sandwell D, Wdowinski S, Frydman S (2002) The lowest place on Earth is subsiding – an InSAR (interferometric synthetic aperture radar) perspective. Geol Soc Am Bull 114:12–23

Bau D, Ferronato M, Gambolati G, Teatini P (2001) Land subsidence spreading factor of the northern Adriatic gas fields, Italy. Int J Geomech 1:459–476

Bertoni W, Bratti C, Carbognin L, Cesi C, Chierici GL, Dossena G, Guerricchio A, La Monica U, La Tegola A, Succetti A (2000) Analysis of subsidence in the Crotone area along the Ionian coast of Calabria, Italy. Proc 6th Int Symp Land Subsid, Ravenna 1:155–166

Borodavko P, Carling P, Parnachov S, Herget J, Clark C, Huggenberger P (2003) The shoreline morphology and the Quaternary lake Chuja-Kuray, Altai mountains, south-central Siberia. XVI INQUA Congress. Geol Soc Am Abstr Programs 183

Brunamonte F, Cavelli S, Serva L, Valente A (2000) Subsidence phenomena in the evolution of Potina Plain, Italy. Proc 6th Int Symp Land Subsid, Ravenna 1:53–58

Bull WB (1973) Geologic factors affecting compaction of deposits in land-subsidence area. Geol Soc Am Bull 84:3783–3802

Carruth RL, Pool DR, Anderson CE (2007) Land subsidence and aquifer-system compaction in the Tucson active management area, south-central Arizona, 1987–2005. US Geol Surv Sci Investig Rep 2007–5190, 27 pp

Castiglia PJ, Fawcett PJ (2006) Large Holocene lakes and climate change in the Chihuahuan Desert. Geology 34:113–116

Chan AW, Zoback M D (2007) The role of hydrocarbon production on land subsidence and fault reactivation in the Louisiana coastal zone. J Coast Res 23:771–786

Chen C, Pei S, Jiao JJ (2003) Land subsidence caused by groundwater exploitation in Suzhou City, China. Hydrogeol J 11:275–287

Chilingar GV, Sorokhtin OG, Khilyuk L, Gorfunkel MV (2009) Greenhouse gases and greenhouse effect. Environ Geol 58:1207–1213

Collier RELL (1990) Eustatic and tectonic controls upon quaternary coastal sedimentation in the Corinth Basin, Greece. J Geol Soc Lond 147:301–314

Colman SM (1998) Water-level changes in Lake Baikal, Siberia: Tectonism versus climate. Geology 26:531–534

Cucci L (2004) Raised marine terraces in the Northern Calabrian Arc (Southern Italy): a – 600 kyr-long geological record of regional uplift. Ann Geophys 47:1391–1406

DeVogel SB, Magee JW, Manley WF, Miller GH (2004) A GIS-based reconstruction of late quaternary paleohydrology: Lake Eyre, arid central Australia. Palaeogr Palaeoclimatol Palaeoecol 204:1–13

Dintelmann D (2004) Comparing the potentiometric surface of a cone of depression near Springfield, Missouri: 1980's and present. Geol Soc Am Abstr Programs 3:3

Drake N, Bristow C (2006) Shorelines in the Sahara:geomorphological evidence for an enhanced monsoon from palaeolake Megachad. Holocene 16:901–911

Dumas B, Gueremy P, Lhenaff R, Raffy J (2006) Rapid uplift, stepped marine terraces and raised shorelines on the Calabrian coast of Messina Strait, Italy. Earth Surf Process Landf 18:241–256

Fairbanks RG (1989) A 17,000-year glacio-eustatic sea level record: influence of glacial melting dates on the Younger Dryas event and deep ocean circulation. Nature 342:637–642

Fjeldskaar W, Cathles L (1991) The present rate of uplift of Fennoscandia implies a low-viscosity asthenosphere. Terra Nova 3:393–400

Fjeldskaar W, Lindholm C, Dehls JF, Fjeldskaar I (2000) Postglacial uplift, neotectonics and seismicity in Fennoscandia. Quat Sci Rev 19:1413–1422

Fleming K, Johnson P, Zwart D, Yokoyama Y, Lambeck K and Chappell J (1998) Refining the eustatic sea-level curve since the last glacial maximum using far- and intermediate-field sites. Earth Planet Sci Lett 163:327–342

Galloway DL, Hudnut KW, Ingebritsen SE, Phillips SP, Peltzer G, Rogez F, Rosen PA (1998) Detection of aquifer system compaction and land subsidence using interferometric synthetic aperture radar, Antelope Valley, Mojave Desert, California. Water Resour Res 34:2573–2585

Gambolati G, Teatini P (2000) The impact of climate change, sea-storm events and land subsidence in the Adriatic (December 1999). FEEM Working Paper No. 21

Garcin Y, Junginger A, Trauth MH, Melnick D, Strecker MR, Olago DO, Maslin M (2008) Major climatic changes in equatorial East Africa during the late Pleistocene and Holocene: reconstruction from paleo-shorelines in the Suguta Valley, northern Kenya Rift. Geophys Res Abstr EGU2008-A-03297

Glowasca E, Sarychikhina O, Suarez F, Nava FA, Mellors R (2010) Anthropogenic subsidence in the Mexicali Valley, Baja California, Mexico, and slip on the Saltillo fault. Environ Earth Sci 59:1515–1524

Gregersen S (2006) Intraplate earthquakes in Scandinavia and Greenland neotectonics or postglacial uplift. J Indian Geophys Union 10:25–30

Griffin DL (2002) Aridity and humidity: two aspects of the late Miocene climate of North Africa and the Mediterranean. Palaeogeogr Palaeoclimatol Palaeoecol 182:65–91

Haq BU, Hardenbol J, Vail P (1987) Chronology of fluctuating sea levels since Triassic (250 million years ago to present). Science 235:1156–1167

Helfert M, Holz R (1985) Multi-source verification of the desiccation of Lake Chad, Africa. Adv Space Res 5:379–384

Hoffman, PF, Kaufman, AJ, Halverson GP, Schrag DP (1998) A neoproterozoic snowball earth. Science 281:1342–1346

Houtenbos APEM. (2000) The quantification of subsidence due to gas-extraction in the Netherlands. Proc 6th Int Symp Land Subsid, Ravenna 1:177–189

IPCC Fourth Assessment Report (2007) Chapter 3 http://www.ipcc.ch/pdf/assessment-report/ar4/wg1/ar4-wg1-chapter3

Jianren L, Lowenstein TK, Brown BC, Ku T-L, Luo S (1996) A 100 ka record of water tables and paleoclimates from salt cores, Death Valley, California. Palaeogr Palaeoclimatol Palaeoecol 123:179–203

Kafri U, Kaufman A, Magaritz M (1983) The rate of Pleistocene subsidence and sedimentation in the Hula Basin as compared with those of other time spans in other Israeli tectonic regions. Earth Planet Sci Lett 65:126–132

Kafri U, Ecker A (1964) Neogene and quaternary subsurface geology and hydrogeology of the Zevulun Plain. Isr Geol Surv Bull 37:1–13

Kolomiets VL (2008) Paleogeography and quaternary terrace sediments and complexes, intermontane basins of Prebaikalia (Southeastern Siberia, Russia). Quat Int 179:58–63

Kominz MA, Miller KG and Browning JV (1998) Long-term and short-term global Cenozoic sea-level estimates. Geology 26:311–314

Kooi H, de Vries JJ (1998) Land subsidence and hydrodynamic compaction of sedimentary basins. Hydrol Earth Syst Sci 2:159–171

Kooi H, Groen J (2003) Geological processes and the management of groundwater resources in coastal areas. Geol Mijnb 82:31–40

Lambeck K (1991) Glacial rebound and sea-level change in the British Isles. Terra Nova 3:379–389

Larson KJ, Basagaoglu H, Marino MA (2001) Prediction of optimal safe ground water yield and land subsidence in the Los Banos-Kettleman City area, California, using a calibrated numerical simulation model. J Hydrol 242:79–102

Leblanc M, Favreau G, Tweed S, Leduc C, Razack M, Mofor L (2007) Remote sensing for groundwater modeling in large semiarid areas: Lake Chad Basin, Africa. Hydrogeol J 15: 97–100

Liu Y, Helm DC (2008) Inverse procedure for calibrating parameters that control land subsidence caused by subsurface fluid withdrawal:1. Methods. Water Resour Res 44:W07423

Liu C, Yu J, Kendy E (2001) Groundwater exploitation and its impact on the environment in the North China Plain. Water Int 26:265–272

Lowenstein TK, Jianren L, Brown C, Roberts SM, Ku T-L, Luo S, Yang W (1999) 200 k.y. paleoclimate record from Death Valley salt core. Geology 27:3–6

Ma R, Wang Y, Ma T, Sun Z, Yan S (2006) The effect of stratigraphic heterogeneity on a real distribution of land subsidence at Taiyuan, northern China. Environ Geol 50:551–568

Magee JW, Miller GH, Spooner NA, Questiaux D (2004) Continuous 150 k.y. monsoon record from Lake Eyre, Australia: insolation-forcing implications and unexpected Holocene failure. Geology 32:885–888

Meehl GA, Washington WM, Collins WD, Arblaster JM, Hu A, Buja LA, Strand WG, Teng H (2005) How much more global warming and sea level rise? Science 307:1769–1772

Miner RE, Nelson ST, Tingey DG, Murrell MT (2007) Using fossil spring deposits in the Death Valley region, USA to evaluate palaeoflowpaths. J Quat Sci 22:373–386

Moerner N-A (1991) Course and origin of the Fennoscandian uplift: the case for two separate mechanisms. Terra Nova 3:408–413

Monjoie A, Paepe R, Su HY (1992) Land subsidence in Shanghai (P. R. of China). Bull Eng Geol Environ 46:5–7

Petalas C, Lambrakis N (2006) Simulation of intense salinization phenomena in coastal aquifer-the case of the coastal aquifers of Thrace. J Hydrol 324:51–64

Popov SV, Shcherba IG, Llyina LB, Nevesskaya LA, Paramonova NP, Khondkarian SO, Magyar I (2006) Late Miocene to Pliocene palaeogeography of the Paratethys and its relation to the Mediterranean. Palaeogeogr Palaeoclimatol Palaeoecol 238:91–106

Qin B, Yu G (1998) Implications of lake levels variations at 6 ka and 18 ka in mainland Asia. Glob Planet Change 18:59–72

Ramirez-Herrera M-T, Urrutia-Fucugauchi J (1999) Morphotectonic zones along the coast of the Pacific continental margin, southern Mexico. Geomorphology 28:237–250

Rieke HH, Chilingar GV (1974) Compaction of argillaceous sediments, developments in sedimentology 16. Elsevier Scientific Publisher, Amsterdam, New York

Rossi A, Calore C, Pizzi U (2000) Land subsidence of Pisa Plain, Italy: Experimental results and preliminary modeling study. Proc 6th Int Symp Land Subsid, Ravenna, 1:91–103

Rostami K, Peltier WR, Mangini A (2000) Quaternary marine terraces, sea-level changes and uplift history of Patagonia, Argentina: comparisons with predictions of the ICE-4G (VM2) model of the global process of glacial isostatic adjustment. Quat Sci Rev 19:1495–1525

Rychagov GI (1997) Holocene oscillations of the Caspian Sea, and forecasts based on palaeogeographical reconstructions. Quat Int 41–42:167–172

Sahagian DL and Holland SM (1991) Eustatic sea-level curve based on a stable frame of reference: preliminary results. Geology 19:1209–1212
Schilling KE (2007) Water table fluctuations under three riparian land covers, Iowa (USA). Hydrol Process 21:2415–2424
Scholz CA, Rosendahl BR (1988) Low lake stands in Lakes Malawi and Tanganayika, East Africa, delineated with multifold seismic data. Science 240:1645–1648
Schmorak S, Mercado A (1969) Upcoming of freshwater–seawater interface below pumping wells. Field study. Water Resour Res 5:1290–1311
Shearer TR (1998) A numerical model to calculate land subsidence, applied at Hangu in China. Eng Geol 49:85–93
Sherif MM, Singh VP (1999) Effect of climate change on sea water intrusion in coastal aquifers. Hydrol Processes 13:1277–1287
Shi X, Wu J, Ye S, Zhang Y, Xue Y, Wei Z, Li Q, Yu J (2008) Regional land subsidence simulation in Su-Xi-Chang area and Shanghai City, China. Eng Geol 100:27–42
Stein M (2001) The sedimentary and geochemical record of Neogene-quaternary water bodies in the Dead Sea basin-inferences for the regional paleoclimatic history. J Paleolimnol 26:271–282
Steinich B, Escolero O, Marin LE (1998) Salt-water intrusion and nitrate contamination in the Valley of Hermosillo and El Sahuaral coastal aquifers, Sonora, Mexico. Hydrogeol J 6:518–526
Sun H, Grandstaff D, Shagam R (1999) Land subsidence due to groundwater withdrawal: potential damage of subsidence and sea level rise in southern New Jersey, USA. Environ Geol 37:290–296
Teatini P, Ferronato M, Gambolati G, Gonella M (2006), Groundwater pumping and land subsidence in the Emilia-Romagna coastland, Italy: modeling the past occurrence and the future trend. Water Resour Res 42:W01406
Terzaghi K (1925) Principles of soil mechanics, IV-settlement and consolidation of clay. Eng News Rec 95:874–878
Trauth MH, Deino AL, Brgner AGN, Strecker MR (2003) East African climate change and orbital forcing during the last 175 kyr BP. Earth Planet Sci Lett 206:297–313
Waldmann N, Stein M, Ariztegui D, Starinsky A (2009) Stratigraphy, depositional environments and level reconstruction of the last interglacial Lake Samra in the Dead Sea basin. Quat Res 72:1–15
Warrick RA, Oerlemans J, Woodworth PL, Meier MF, Le Provost C (1996) Changes in sea level. In: Houghton JT, Meira Filho LG, Callander BA (eds) Climate Changes 1995: The Science of Climate, Contribution of Working Group I to the Second Assessment Report of the Intergovernmental Panel of Climate Changes. Cambridge University Press, 359–405
Williams PW (1982) Speleothems dates, quaternary terraces and uplift rates in New Zealand. Nature 298:257–260
Winter TC (1999) Relation of streams, lakes, and wetlands to groundwater flow systems. Hydrogeol J 7:28–45
Winter TC, Harvey, JW, Franke OL, Alley WM (2002) Groundwater and surface water – a single resource. US Geol Surv Circular 1139:79
Yang B, Shi Y, Braeuning A, Wang J (2004) Evidence for a warm-humid climate in arid northwestern China during 40–30 ka BP. Quat Sci Rev 23:2537–2548
Zhang Y, Xue Y-Q, Wu J-C, Yu J, Wei Z-X, Li Q-F (2007) Land subsidence and earth fissures due to groundwater withdrawal in the southern Yangtze Delta, China. Environ Geol 55:751–762
Zhou X, Yan X, Li J, Yao J, Dai W (2007) Evolution of the groundwater environment under a long-term exploitation in the coastal area near Zhanjiang, China. Environ Geol 51:847–856

Chapter 4
Methods and Techniques to Define Base-Level Elevation and to Measure and Assess the Effect of Their Variation on Adjoining Groundwater Systems

This chapter describes the different methods employed to detect, measure, monitor and estimate current and paleo base-levels in time and space. These include ground elevation as well as sea and lake levels. These methods include direct measurement, such as geodetic methods and level gauges and indirect methods, using proxies such as sedimentological, morphological and chemical data for past records. The methods of measuring the response of groundwater system to the base-level changes are also described.

4.1 Base-Level Elevation

4.1.1 Current Ground Level Measurements

Current ground levels and their changes are instrumentally measured and monitored by repeated geodetic measurements employing the following techniques:

4.1.1.1 Precise Geodetic Leveling

Repeated precise geodetic levelings are used to define elevations and to monitor current vertical movements of uplift and subsidence. The first order precise leveling method (Rappleye 1948) was often used to study Recent Crustal Movements attaining a precision of a few millimeters. The usage of the above was reported, for example, from Israel by Kafri (1969) and Karcz and Kafri (1971, 1973). Vertical deformation in the Main Ethiopian Rift was also recorded employing repeated leveling (Asfaw et al. 2006). Precise leveling of geodetic networks can thus be a helpful tool to monitor current elevation changes of continental base-levels.

U. Kafri and Y. Yechieli, *Groundwater Base Level Changes and Adjoining Hydrological Systems*,
DOI 10.1007/978-3-642-13944-4_4, © Springer-Verlag Berlin Heidelberg 2010

4.1.1.2 The GPS Technique

The Global Positioning System (GPS) is a system developed by the U.S. Department of Defense. It uses a constellation of between 24 and 32 earth orbiting satellites that transmit precise signals to receivers on earth thereby enabling to determine their location and elevation. The basics of the method are described for example by Vanicek and Krakiwsky (1986). The vertical precision of the system is around a few millimeters.

GPS measurements are used to define elevations. Repeated GPS network measurements enable to detect current vertical elevation changes of uplift and subsidence on land. Examples to such studies were described from Greenland (Forsberg et al. 2000), from Tibet (Xu et al. 2000) and from Turkey (Ustun and Demirel 2006). Thus, The GPS technique can serve as a useful tool to detect on-land vertical base-level changes.

4.1.1.3 The INSAR Technique

The Interferometric Synthetic Aperture Radar (INSAR) technique has become a valuable technique to measure displacement at the ground surface (i.e., Gabriel et al. 1989; Massonnet and Feigl 1998). When two radar scans are made at different times from the same viewing angle, a small change in the position of the target, namely the ground surface, may create a detectable change in the phase of the reflected signals. The resulting phase differences are expressed in interferograms in which the fringe pattern reflects the ground displacement that occurred between the two acquisitions.

Reported studies that used the INSAR technique to measure land subsidence are, among others, from California (Galloway et al. 1998) from the Dead Sea (Baer et al. 2002) and in the Asal Rift, Djibouti (Doubre and Peltzer 2007).

4.1.2 Current Sea and Lake Level Measurements

The instrumental technique to measure and monitor current sea and lake levels is by using tide gauges (i.e., Douglas et al. 2001). The current global sea level rise due to the on-going global warming process is being an important issue and thus being detected by tide gauges from all over the world (i.e., Cazenave et al. 1999; Miller and Douglas 2003; Holgate et al. 2008; Blasi 2009). In addition, Satellite Radar Altimetry is also often used to monitor lake level changes attaining a precision of a few centimeters (i.e., Birkett 1995; Cretaux and Birkett 2006).

Significant declines of lake levels in the last decades, using the above techniques, were reported among others from the Caspian Sea (Cazenave et al. 1997), from Lake Chad (Birkett 2000), from the Aral Sea (Peneva et al. 2004) and from the Dead Sea (Yechieli et al. 2006).

4.1.3 *Paleo- and Historic Shorelines*

The indicators that enable reconstruction of paleo-shorelines include near-shore marine terraces, wave-cut platforms and notches, near-shore cave formation, submerged archeological sites, geomorphological features, as well as sedimentological evidence for near-shore marine or lacustrine environments. Geochemical and isotopic data are able to reveal the specific (e.g., salinity, temperature, water levels) conditions of the ancient environments. All the above evidences are often found above or below the present base-level due to tectonic vertical displacements as well as global or local climatic changes which may lead to desiccation of lakes.

4.1.3.1 Near-Shore Marine Terraces

Near-shore marine terraces are basically divided into two main types, namely the constructional type and the erosional one. The constructional type consists of reef platforms which typify the shallow near-shore zone. The erosional type consists of wave-cut platforms or abrasion ramps as well as wave-cut notches that are formed by destructive wave action along shore lines and against the base of sea cliffs. These two phenomena are described and exhibited (Fig. 4.1) by Burbank and Anderson (2000).

Thus, ancient wave-cut platforms and notches are indicative of ancient shore lines and provide evidence of past eustatic sea level changes. Raised and abandoned platforms and notches as well as coastal sediments and reefs are evidences of higher past sea levels. Such were described from Australia (Cooke 1971), from the Cayman Islands (Jones and Hunter 1990) and from Barbados (Johnson 2001).

Ancient shore lines phenomena were also often subjected to later tectonic displacements both uplift and subsidence. Examples are described from Mexico (Ramirez-Herrera and Urrutia-Fucugauchi 1999), from Italy (Cucci 2004; Dumas

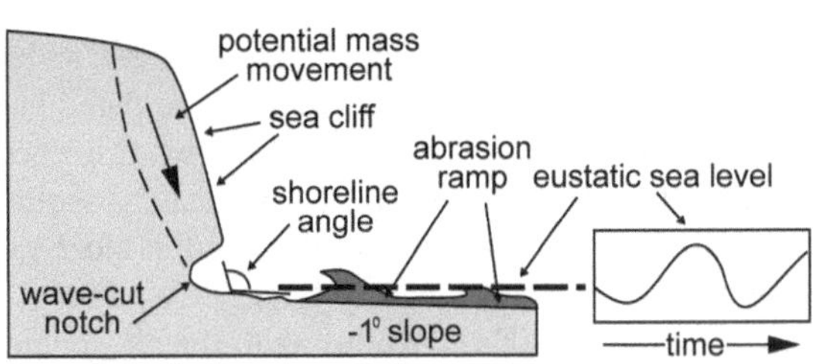

The vertical position of the abrasion ramp varies through times as a function of sea-level variations

Fig. 4.1 Schematic description of a near shore abrasion ramp and a wave-cut notch [after Burbank and Anderson (2000)]

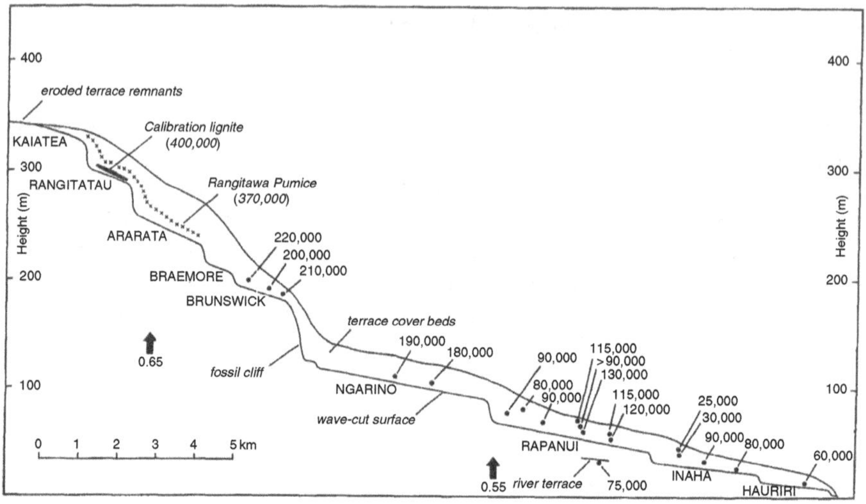

Fig. 4.2 Tectonically uplifted ancient shore terraces in New Zealand. Uplift rates in millimeter/year are shown by *large arrows* [after Pillans (1983)]

et al. 2006), from Spain (Alvarez-Marron et al. 2008), from Argentina (Rostami et al. 2000) and from New Zealand (Pillans 1983) (Fig. 4.2). Raised shorelines due to glacial isostatic uplift in Fennoscandia were also described, among others by Moerner (1991) (see also Sect. 3.4).

The slope of the surfaces of near-shore terraces can serve as a tool to assess a down-stream lake level, since the evolution of the alluvial fans surfaces and the lakes are interrelated. Thus, an approximate record the lake level at different times is obtained (Bowman 1971; Sneh 1979).

4.1.3.2 Karstification and Cave Formation

The delineation of paleokarstic features in carbonate aquifers, and mostly horizontal cave levels, is a powerful tool to allocate and interpret paleo base-levels.

Horizontal or sub-horizontal cave systems, which are not formed by hydrothermal processes, are usually related to enhanced carbonate dissolution due to fast horizontal flow in the shallow phreatic zone of the aquifer, close to the water table (i.e., Swinnerton 1932; Rhoades and Sinacori 1941; Davis 1960; Thrailkill 1968; Esteban 1987; Hill 1999; Flugel 2004). According to the above, enhanced dissolution and cave formation (speleogenesis) in the shallow phreatic zone, takes place in cases of long period of stable water table, and rapid groundwater flow. Thus, a cave system level, whether above the current groundwater table and sea level or submerged below it, might reflect a paleo-groundwater level. When extending these levels using reasonable water table gradients, the location and elevation of the adjoining paleo base-level can be obtained.

High groundwater fluxes are especially evident in an aquifer adjacent to its base-level where the latter is marine or saline lake. In such a case, an interface between the overlying flowing fresh groundwater and the underlying saline water body exists. As a result, the fresh groundwater flow converges, close to the base-level to a very narrow discharge zone accompanied by rapid fluxes (Fig. 4.3). Such cases were described from the Dead Sea coastal aquifer by Yechieli et al. (2001) and Kafri and Yechieli (2010) and from North Carolina by Evans and Lizarralde (2003). In the case of hypersaline lakes, the interface is shallower than in the case of normal seawater, and the overlying groundwater velocity therefore is more rapid (see also Sect. 6.2.2), which might enhance cave formation.

Another factor which controls carbonate dissolution and cave formation in the near-shore of coastal aquifers is the mixing zone between fresh and saline waters. Cave formation related to the mixing zone were described, among others, from Yucatan by Back et al. (1986), from south Pacific by Murgulet and Aharon (2005) and in relation to carbonate islands by Mylroie and Carew (1995) and Jenson et al. (2006).

Paleokarstic features related to paleo-water tables and/or base-levels are often found at different levels above the present base levels or submerged below them as a result of global eustatic sea level changes, drop or rise of lake levels due to global

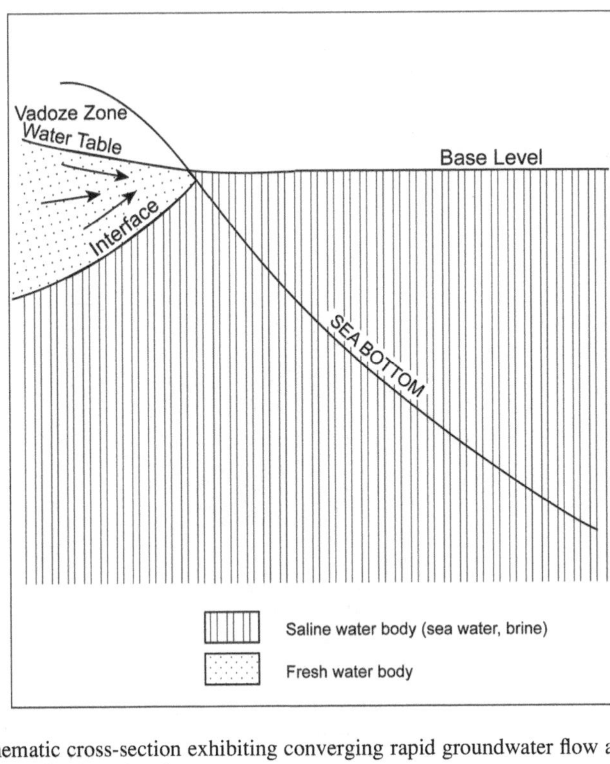

Fig. 4.3 Schematic cross-section exhibiting converging rapid groundwater flow above the interface, close to a saline base level

or local climate changes, and vertical tectonic displacement. The age of the formation of such paleokarstic cave systems can be obtained by the following:

(a) Absolute elevation of a cave system as compared to known eustatic elevation curves, provided the age of the eustatic level is known.
(b) When associated with nearby near shore sediments, beach terraces or coral reefs, paleontological or absolute dating might yield the date of the caves (i.e., Williams 1982; Florea et al. 2007) (Fig. 4.4). Dated lacustrine sediments, which were deposited within an already existing cave adjacent to the shore line, can also yield the youngest age threshold of the cave as shown by Lisker et al. (2009).
(c) Absolute dating of speleothems in caves yields a minimal age of cave formation since speleothems are formed only in air filled caves in the vadose zone following their emergence above the groundwater table (i.e., Atkinson et al. 1978).

Caves ceased to develop when they were subsequently submerged below the rising sea levels. Examples of raised and submerged, often multiple, caves systems and related coastal terraces were described among others by Gascoyne et al. (1979), Williams (1982), Li et al. (1989), Collier (1990), Richards et al. (1994), Carew and Mylroie (1995), Jenson et al. (2006) and Florea et al. (2007).

Data which might support paleo base-level delineation and interpretation becomes more abundant when younger base-levels are studied, partly due to a better dating resolution (e.g., radiocarbon dating) and partly to better preservation. The Neogene period already provides some information but considerably more information is available regarding Quaternary paleo-karst and base-levels.

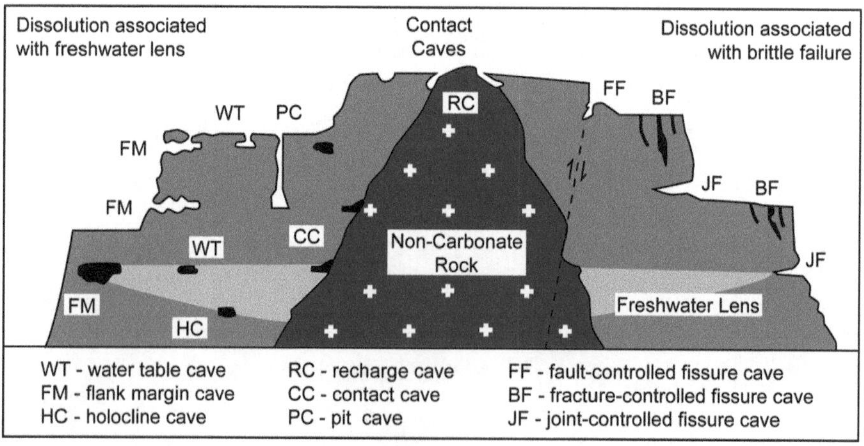

Fig. 4.4 Uplifted cave levels in the Mariana Islands [after Jenson et al. (2006)]

4.1.3.3 Erosional and Other Geomorphological Evidence

Among the erosional and geomorphological features which are indicative of past base-level elevations are coastal abrasion platforms and lake shore terraces (Figs. 4.5 and 4.6) (i.e., Bowman 1971; Kafri and Arad 1978). All the above listed features are described in detail, among others, by Reading (1996) and Cohen (2003).

Receding paleo-lake shorelines are also detected by analysis of longitudinal profiles of streams that drained to those lakes. Receding shorelines in the Dead Sea Basin are manifested in adjoining longitudinal stream profiles as well as by a sequential development of alluvial fans (Bowman et al. 2007).

Fig. 4.5 Dead Sea coastal abrasion lake shore terraces

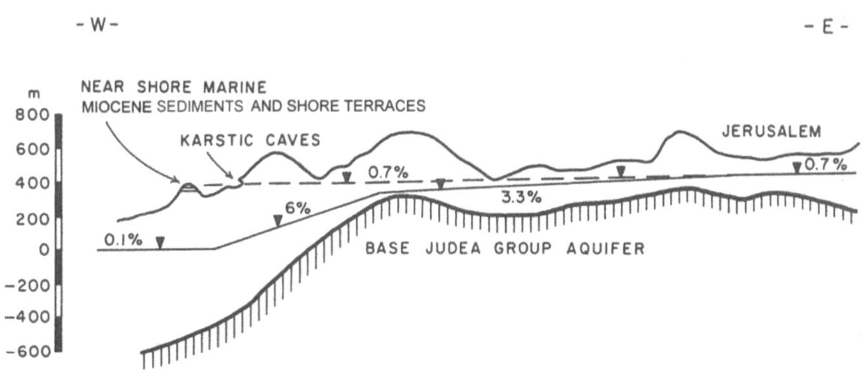

Fig. 4.6 Tectonically raised Neogene abrasion plain and related paleohydrological system [after Kafri and Arad (1978)]. The *solid line* represent the present water table connected to the present Mediterranean base level. The *dashed line* represent the reconstructed paleo water table related to the Miocene base level

4.1.3.4 Sedimentological Indicators

The role of different sediments as indicators of the environment of deposition was discussed in numerous books and articles (i.e., Nicols 1999).

Among the sedimentological indicators are lacustrine sands and laminated sediments, deltaic sediments, beach ridges, alluvial fans as well as evaporites in cases of saline lakes. Some sediments, such as beach ridges, indicate the exact location of the shoreline (Bartov et al. 2006; Bookman et al. 2004) while others indicate minimal or maximal water level. Lacustrine sediment (e.g., aragonite, gypsum, halite) indicate the existence of a lake whereas alluvial sediments (e.g., gravel and conglomerate) imply that the lake did not exist there at that time. An exception could be in case of submarine erosion, depositing detrital material offshore in the lake, at the mouth of big rivers.

Paleontological or pollen studies of lake sediment yield information regarding lake and/or sea water depth and salinity (e.g., Migowski et al. 2006; Begin et al. 1980). Geochemical ratios and stable isotope values (e.g., Na/Cl, $\delta^{18}O$) also serve as good proxies for the determination of the specific salinity and temperature conditions of ancient environments (i.e., Katz et al. 1977). Thus, a water level profile can be obtained from sedimentological sequences.

The reconstruction of paleo-shorelines and lake levels from the sedimentological data requires dating of these sediments. The most commonly used dating method is that of radiocarbon, employed for carbonates such as aragonite or calcite minerals. The ^{14}C radioisotope has a half life of 5,700 years and thus the method enables dating of material up to ~40,000 years old. Other dating methods include the U-Th method, mostly for carbonates (e.g., Kaufman and Broecker 1965) and the OSL method (Aitken 1998) which can be applied to a variety of sediments (e.g., quartz grains) and to sediments older than 500,000 years.

Dated Quaternary shorelines were described from numerous places such as Lake Van, Turkey (Landmann et al. 1996), from Lake Baikal (Colman 1998; Kolomiets 2008), from the Great Lakes, North America (Baedke and Thompson 2000; Baedke et al. 2004) and from the Dead Sea, Israel (Bookman et al. 2004; Waldmann et al. 2007).

4.1.3.5 Submerged Archaeological Sites

Some coastal archaeological sites are found at present to be submerged below sea level due to the rise of sea level since the Holocene. The dating of those sites enables to reconstruct and to assess the rate of the sea level rise in the historic and pre-historic time span.

Submerged archaeological sites and human installations were described, among others, from offshore northern Israel (Galili and Nir 1993; Galili and Sharvit 1998), from the Black Sea (Coleman and Ballard 2007) and from Italy (Scicchitano et al. 2008).

4.2 Methods to Determine Groundwater Systems' Response to Base-Level Changes

Studies concerning groundwater response to base-level changes, which employ field measurements, are still rare. The following chapter, therefore, describes mainly methods which were employed in the Dead Sea area.

4.2.1 Current Field Measurement

The response of the groundwater system to changes in base-level elevation is expected to be expressed in the field by the following: groundwater level changes, vertical and lateral shift of the fresh–saline water interface in the case of a saline base-level and changes in submarine or sublacustrine groundwater discharge (SGD). Water level measurements and monitoring are carried out employing manual meters, electric devices or pressure transducers at different resolutions and frequencies, depending on the specific objective. An example of groundwater level response to the Dead Sea level changes was described by Yechieli et al. (1995, 2009b) and Kiro et al. (2008) (see also Sect. 11.2.6.3).

The electrical conductivity (EC) profiling and logging of groundwater in boreholes is a known method to detect the fresh–saline water interface. In general, at salinity of up to normal marine values, the EC value correlates satisfactorily with salinity values. The situation is somewhat more complicated at higher salinities but still correlation exists up to a salinity of half of the Dead Sea brine. At a higher salinity, no correlation is found between EC and salinity (Yechieli 2000). The EC profiles are repeatedly conducted for several years in order to examine the response of the interface to changes of base-level elevation. A significant drop of the interface due to changes in base-level elevation was indeed observed in the Dead Sea system (Yechieli 2000; Kiro et al. 2008) (see also Sect. 11.2.6.3). Continuous EC monitoring is also carried out in order to study short term processes. However, the interpretation should be done with cautious due to both artificial effects of the borehole itself and to the tidal effect (e.g., Shalev et al. 2009).

The base-level changes can also be manifested in the amount of submarine groundwater discharge (SGD) to the sea or to a lake. Such changes are expected since a drop of the base-level changes the hydraulic gradient near the shoreline and thus an increase in groundwater flow is expected partly on the expense of the storage. The SGD is estimated by direct measurement with seepage meters (Taniguchi and Iwakawa 2001) which may represent the local situation. An additional way is by sampling and analyzing the seeping water for chemical and isotopic analysis, based on the fact that they differ from the surrounding water body and thus the obtained results are of a more regional significance. The most commonly used measured constituents, in recent studies, are radon (Burnett and Dulaiova 2003) and radium isotopes (Moore 1996), although other parameters, such as nutrients and various pollutants are also used.

The increase of water discharge to the base-level can also be measured on land by in situ velocity meters which are installed in exploration boreholes. Such a device can monitor the change of groundwater flow velocity and direction of both fresh and saline groundwater. A complementary study could be done with artificial tracers such as dyes (e.g., rhodamine) using either the point dilution test, if only one borehole exists, or preferably using an array of monitoring wells if available. Tracers should be chosen with caution as to match the specific requirements, i.e., high salinity (Magal et al. 2008).

4.2.2 Indirect Estimation

Chemical concentrations and ratios, as well as both stable and radioactive isotopes in groundwater, can be used to estimate the groundwater regime between the intake area and the discharge zone and thus its response to base-level changes. Stable isotopes such as ^{18}O, together with noble gas information (Stute et al. 1992), yield an insight as to the temperature that prevailed in the recharge zone, and thus climate and/or altitude of the intake area.

The radio isotopic methods are used, together with the above methods, to determine the age of the groundwater, at certain points along the flow path, from the recharge zone to the discharge area, and thus its flow velocity and residence time. Several radio isotopic methods are used, each with different time scale, depending on their half life time. A general description of these isotopes is given by Phillips and Castro (2003). The most commonly used radio isotopes for groundwater dating are radiocarbon (half life time of ~5,700 years, Munnich 1957) and tritium (half life time of 12.4 years, Bergman and Libby 1957). The radiocarbon method, with the more extended time window, requires several corrections and modification before interpretation, due to its interaction with the rock matrix, especially in carbonate aquifers.

There are several methods of corrections, depending on the specific conditions of the aquifers, some of which are computer aided codes (e.g., Netpath, Plummer et al. 1991). The advantage of tritium is its being part of the water molecule and acting as a conservative parameter. On the other hand, its relative short half life time limits its use to the current short term time window processes. For several decades, the tritium signal was significantly larger due to the fallout of the nuclear testing and therefore a useful tool for groundwater dating. The decreasing tritium values in recent years, due to the cessation of the nuclear tests, limit its usage. Tritium can be, thus, best used together with the analysis of ^{3}He to provide reliable ages of young groundwater (Ekwurzel et al. 1994).

Other less common methods of groundwater dating include several radioisotopes such as ^{36}Cl (half life time of 300,000 years), Noble gases (e.g., ^{39}Ar with half life time of 269 years, Loosli 1983; Loosli et al. 2000), ^{85}Kr (with half life time of

10.76 years), ^{81}Kr (with half life of 229,000 years, Loosli and Oeschger 1969; Lehmann et al. 1991). Depending on the specific half life of each isotope, they are applicable for different processes of different time scales. However, these isotopes were seldom used since their analytical procedure is less available and relatively complicated. The application of the accumulation rate of ^{4}He was suggested to be a qualitative dating tool of old groundwater (Patterson et al. 2005). The Rn content (half life of 3.8 day) can also provide, in several cases, information with regard to groundwater flow rates (i.e., Kafri 2001).

Other methods of groundwater age determination include several pollutants whose concentration in the atmosphere have increased in the twentieth century and are relatively well known (e.g., SF_6, CFC, Busenberg and Plummer 1992). These are, thus, applicable for dating of modern groundwater, younger than 70 years.

Dating of fresh and saline groundwater indicate the travel time of water from the recharge area to the sampling point (monitoring boreholes or discharging springs) or from the sea or lake into the aquifer. A change in base-level is expected to cause a change in travel time of fresh groundwater which may be detected in some systems. Dating of saline groundwater in coastal aquifers indicates the timing of inland seawater or saline lake water intrusion which, in turn, implies on the timing of the sea or lake level rise (i.e., Yechieli et al. 2009a).

The response of coastal aquifers waters to variations in base-level can also be manifested by changes in their chemical composition and ratios, as well as by their isotopic composition, due to changes in the contribution of the different end members to the groundwater. Such changes were attributed, for example, in the receding Dead Sea system to flushing of the brine from the aquifers or to dissolution of evaporites (Yechieli 2006; Kiro et al. 2008).

Chemical changes could be also manifested in the ionic ratio (e.g., Ca/SO_4, Na/Cl) due to dissolution or precipitation of minerals, such as gypsum or halite.

4.2.3 Hydrological Simulations

Hydrological simulations also serve as a tool to analyze the response of all the aspects of the hydrological system to base-level changes by using existing codes such as SUTRA (Voss 1984), SEAWAT (Langevin et al. 2008) and FEFLOW (Diersch and Kolditz 2002). These codes take into account the effect of density driven flows which are important in the case of systems adjacent to saline base-levels. The advantage of hydrological simulations is that they have no time limitation and can be extended to long periods for past situations as well as for future forecast. The simulation can be used to complete missing past records and for forecasting future situation. The problematics of hydrological simulation is its requirement of several hydraulic properties, such as hydraulic conductivity, that are difficult to obtain and thus the simulations yield, in many cases, only rough estimates.

As always, due to the uncertainty of each separate method, it is preferred to use a combination of several methods in order to achieve more confidence in the results.

References

Aitken MJ (1998) An introduction to optical dating. Oxford Science Publication, New York, 267

Alvarez-Marron J, Hetzel R, Niedermann S, Menendez R, Marquinez J (2008) Origin, structure and exposure history of a wave-cut platform more than 1 Ma in age at the coast of northern Spain: a multiple cosmogenic nuclide approach. Geomorphology 93:316–334

Asfaw LM, Beyene H, Mkonnen A, Oli T (2006) Vertical deformation in the main Ethiopian rift: leveling results in its northern part, 1995–2004. In: Yirgu G, Ebinger CG, Maguire PKH (eds) The afar volcanic province within the East African rift system, Geological Society London Special Publications, London 259:185–190

Atkinson TC, Harmon RS, Smart PS, Waltham AC (1978) Paleoclimatic and geomorphic implications of 230^{Th}/234U dates on speleothems from Britain. Nature 272:24–28

Back W, Hanshaw BB, Herman JS, Van Driel JN (1986) Differential dissolution of a Pleistocene reef in the ground-water mixing zone of coastal Yucatan, Mexico. Geology 14:137–140

Baedke SJ, Thompson TA (2000) A 4,700-year record of lake level and isostasy for lake Michigan. J Great Lakes Res 26:416–426

Baedke SJ, Thompson TA, Johnston JW, Wilcox DA (2004) Reconstructing paleo lake levels from relict shorelines along the upper great lakes. Aquat Ecosyst Health Manag 7:435–449

Baer G, Schattner U, Wachs D, Sandwell D, Wdowinski S, Frydman S (2002) The lowest place on earth is subsiding – an InSAR (interferometric synthetic aperture radar) perspective. Geol Soc Am Bull 114:12–23

Bartov Y, Agnon A, Enzel Y, Stein M (2006) Late quaternary faulting and subsidence in the central Dead Sea basin. Isr J Earth Sci 55:17–31

Begin ZB, Nathan Y, Ehrlich A (1980) Stratigraphy and facies distributions in the Lisan formation: new evidence from the area south of the Dead Sea, Israel. Isr J Earth Sci 29:182–189

Bergman F, Libby WF (1957) Continental water balance, groundwater inventory and storage times, surface ocean mixing rates, and worldwide water circulation patterns from cosmic ray and bomb tritium. Geochim Cosmochim Acta 12:277–296

Birkett CM (1995) Contribution of TOPEX/POSEIDON to the global monitoring of climatically sensitive lakes. J Geophys Res 100:25179–25204

Birkett CM (2000) Synergistic remote sensing of Lake Chad: variability of basin inundation. Rem Sens Environ 72:218–236

Blasi CJ (2009) A new technology for the measurement of the sea level and the sea state. Environ Geol 57:331–336

Bookman Ken-Tor R, Enzel Y, Agnon A, Stein M (2004) Late Holocene lake levels of the Dead Sea. Geol Soc Am Bull 116:555–571

Bowman D (1971) Geomorphology of the shore terraces of the late Pleistocene Lisan Lake (Israel). Palaeogeogr Palaeoclimatol Palaeoecol 9:183–209

Bowman D, Shachnovich-Firtel Y, Devora S (2007) Stream channel convexity induced by continuous base-level lowering, the Dead Sea, Israel. Geomorphology 92:60–75

Burbank DW, Anderson RS (2000) Tectonic geomorphology. Blackwell Publishing, London

Burnett WC, Dulaiova H (2003) Estimating the dynamics of groundwater input into the coastal zone via continuous Radon-222 measurements. J Environ Radioact 69:21–35

Busenberg E, Plummer, N (1992) Use of chlorofluorocarbons (CCl_3F and CCl_2F_2) as hydrological tracers and age dating tools: the alluvial and terrace system of central Oklahoma. Water Resour Res 36:3011–3030

Carew JL, Mylroie JE (1995) Quaternary tectonic stability of the Bahamian archipelago: evidence from fossil coral reefs and flank margin caves. Quat Sci Rev 14:145–153

Cazenave A, Bonnefond P, Dominh K, Schaeffer P (1997) Caspian sea level from Topex–Poseidon altimetry: level now falling. Geophys Res Lett 24:881–884

Cazenave A, Dominh K, Ponchaut F, Soudarin L, Cretaux JF, Le Provost C (1999) Sea level changes from Topex–Poseidon altimetry and tide gauges, and vertical crustal motions from DORIS. Geophys Res Lett 26:2077–2080

Cohen AS (2003) Paleolimnology. The history and evolution of lake systems. Oxford University Press, New York

Coleman DF, Ballard RD (2007) Submerged paleoshorelines in the southern and western Black Sea – implications for inundated prehistoric archaeological sites, In: Yanko-Hombach V, Gilbert AS, Panin N, Dolukhanov PM (eds) The Black Sea flood question: changes in coastline, climate, and human settlement. Springer, The Netherlands

Collier RE LL (1990) Eustatic and tectonic controls upon quaternary coastal sedimentation in the Corinth Basin, Greece. J Geol Soc London 147:301–314

Colman S M (1998) Water-level changes in Lake Baikal, Siberia: tectonism versus climate. Geology 26:531–534

Cooke CW (1971) American emerged shorelines compared with levels of Australian marine terraces. Geol Soc Am Bull 82:3231–3234

Cretaux J-F, Birkett C (2006) Lake studies from satellite radar altimetry, C R Geosci 338:1098–1112

Cucci L (2004) Raised marine terraces in the Northern Calabrian Arc (Southern Italy): a – 600 kyr-long geological record of regional uplift. Ann Geophys 47:1391–1406

Davis WE (1960) Origin of caves in folded limestones. Natl Speleol Soc Bull 22:5–18

Diersch H-JG, Kolditz O (2002) Variable-density flow and transport in porous media: approaches and challenges. Adv Water Resour 25:899–944

Doubre C, Peltzer G (2007) Fluid-controlled faulting process in the Asal Rift, Djibouti, from 8 yr of radar interferometry observations. Geology 35:69–72

Douglas BC, Kearney MS, Leatherman SP (2001) Sea level rise. History and consequence. Int Geophys Ser 75:232

Dumas B, Gueremy P, Lhenaff R, Raffy J (2006) Rapid uplift, stepped marine terraces and raised shorelines on the Calabrian coast of Messina Strait, Italy. Earth Surf Process Landf 18:241–256

Ekwurzel B, Schlosser P, Smethie WM Jr, Plummer N, Busenberg E, Michel RL, Weppernig R, Stute M (1994) Dating of shallow groundwater: comparison of the transient tracers $^{3}H/^{3}He$, chlorofluorocarbons and ^{85}Kr. Water Reosur Res 30:1693–1708

Esteban M (1987) Unconformities, paleokarst facies, and porosity evolution. Am Assoc Petrol Geol Bull 71:1440

Evans RL, Lizarralde D (2003) Geophysical evidence for karst formation associated with offshore groundwater transport: an example from North Carolina. Geochem Geophys Geosyst 4:1069

Florea LJ, Vacher HL, Donahue B, Naar D (2007) Quaternary cave levels in peninsular Florida. Quat Sci Rev 26:1344–1361

Flugel E (2004) Microfacies of carbonate rocks. Analysis, interpretation and application. Springer, Berlin

Forsberg R, Keller K, Nielsen CS, Gundestrup N, Tscherning CC, Madsen SN, Dall J (2000) Elevation change measurements of the Greenland ice sheet. Earth Planets Space 52:1049–1053

Gabriel AK, Goldstein RM, Zebker HI (1989) Mapping small elevation changes over large areas: differential radar interferometry. J Geophys Res 94:9183–9191

Galili E, Nir Y (1993) The submerged pre-pottery Neolithic water well of Atlit-Yam, northern Israel, and its palaeoenvironmental implications. Holocene 3:265–270

Galili E, Sharvit J (1998) Ancient coastal installations and the tectonic stability of the Israeli coast in historical times. Geol Soc Lond Spec Publ 146:147–163

Galloway DL, Hudnut KW, Ingebritsen SE, Phillips SP, Peltzer G, Rogez F, Rosen PA (1998) Detection of aquifer system compaction and land subsidence using Interferometric Synthetic Aperture Radar, Antelope Valley, Mojave Desert, California. Water Resour Res 34:2573–2585

Gascoyne M, Benjamin GJ, Schwarcz HP, Ford DC (1979) Sea-level lowering during the Illinoian glaciation: evidence from a Bahama "Blue Hole". Science 205:806–808

Hill CA (1999) Overview of Kartchner caverns, Arizona. J Cave Karst Stud 61:41–43

Holgate S, Foden P, Pugh J, Woodworth P (2008) Real time sea level data transmission from tide gauges for tsunami monitoring and long term sea level rise observations. J Oper Oceanogr 1:3–8

Jenson JW, Keel TM, Mylroie JE, Stafford KW, Taborosi D, Wexel C (2006) Karst of the Mariana Islands: the interaction of tectonics, glacio-eustasy, and freshwater/seawater mixing in island carbonates. Geol Soc Am Spec Pap 404:129–138

Johnson RG (2001) Last interglacial sea stands on Barbados and an early anomalous deglaciation timed by differential uplift. J Geophys Res 106(C6):11,543–11,551

Jones B, Hunter IG (1990) Pleistocene paleogeography and sea levels on the Cayman Islands, British West Indies. Coral Reefs 9:81–91

Kafri U (1969) Recent crustal movements in northern Israel. J Geophys Res 74:4246–4258

Kafri U (2001) Radon in groundwater as a tracer to assess flow velocities: two test cases from Israel. Environ Geol 40:392–398

Kafri U, Arad A (1978) Paleohydrology and migration of the ground-water divide in regions of tectonic instability in Israel. Geol Soc Am Bull 89:1723–1732

Kafri U, Yechieli Y (2010) The role of hydrogeological base level in the formation of sub-horizontal caves horizons, example from the Dead Sea Basin, Israel. Environ Earth Sci DOI:10.1007/s12665-009-0435-4

Karcz I, Kafri U (1971) Recent crustal movements in the Negev, southern Israel. J Geophys Res 76:8056–8065

Karcz I, Kafri U (1973) Recent vertical crustal movements. Nature 242:42–44

Katz A, Kolodny Y, Nissenbaum A. (1977) The geochemical evolution of the Pleistocene Lake Lisan – Dead Sea system. Geochim Cosmochim Acta 41:1609–1626

Kaufman A, Broecker WS (1965) Comparison of ^{230}Th and ^{14}C ages of carbonate materials from Lake Lahontan and Bonneville. J Geophys Res 70:4039–4054

Kiro Y, Yechieli Y, Lyakhovsky V, Shalev E, Starinsky A (2008) Time response of the water table and saltwater transition zone to a base level drop. Water Resour Res 44:W12442, DOI:10.1029/2007WR006752

Kolomiets VL (2008) Paleogeography and quaternary terrace sediments and complexes, intermontane basins of Prebaikalia (Southeastern Siberia, Russia). Quat Int 179:58–63

Landmann G, Reimer A, Kempe S (1996) Climatically induced lake level changes at Lake Van, Turkey, during the Pleistocene/Holocene transition. Global Biogeochem Cycles 10:797–808

Langevin CD, Thorne D, Dausman AM, Sukop MC, Guo W (2008) SEAWAT Version 4: a computer program for simulation of multi-species solute and heat transport. US geological survey techniques and methods book 6, chapter A22, 39p

Lehmann BE, Loosli HH, Rauber D, Thonnard N, Willis RD (1991) ^{81}Kr and ^{85}Kr in groundwater, Milk River Aquifer, Alberta, Canada. Appl Geochem 6:847–868

Li W-X, Lundberg J, Dickin AP, Ford DC, Schwartz HP, McNutt R, Williams D (1989) High-precision mass-spectrometry uranium series dating of cave deposits and implications for palaeoclimate studies. Nature 339:534–536

Lisker S, Vaks A, Bar-Matthews M, Porat R, Frumkin A (2009) Stromatolites in caves of the Dead Sea fault escarpment: implications to latest Pleistocene lake levels and tectonic subsidence. Quat Sci Rev 28:80–92

Loosli HH (1983) A dating method with ^{39}Ar. Earth Planet Sci Lett 63:51–62

Loosli HH, Lehmann BE, Smethie WM Jr (2000) Noble gas radioisotopes: ^{37}A, ^{85}Kr, ^{39}Ar, ^{81}Kr. In: Cook P and Herczeg A (eds) Environmental tracers in subsurface hydrology). Kluwer Academic, London pp 379–396

Loosli HH, Oeschger H (1969) ^{37}Ar and ^{81}Kr in the atmosphere. Earth Planet Sci Lett 7:67–71

Magal E, Weisbrod N, Yakirevitz A, Yechieli Y (2008) Artificial tracer in saline groundwater. J Hydrol 358:124–133

Massonnet D, Feigl K (1998) Radar interferometry and its application to changes in the Earth's surface. Rev Geophys 36:441–500

Migowski C, Stein M, Prasad S, Agnon A, Negendank JFW(2006) Dead Sea levels, climate variability and human culture evolution in the Holocene Near East. Quat Res 66:421–431

Miller L, Douglas BC (2003) Sea level rise during the past 80 years determined from tide gauge and hydro observations: do they agree? Geophys Res Abstr 5:14548

Moerner N-A (1991) Course and origin of the Fennoscandian uplift: the case for two separate mechanisms. Terra Nova 3:408–413

Moore WS (1996) Large groundwater inputs to coastal waters revealed by ^{226}Ra enrichments. Nature 380: 612–614

Munnich KO (1957) Messungen des ^{14}C Gehaltes vom hartem grundwasser. Naturwissenschaften 44:32–33

Murgulet V, Aharon P (2005) A groundwater chemical drill explains caves formation on Niue Island, south Pacific. Geol Soc Am Abstr Programs 37:40

Mylroie JE, Carew JL (1995) Karst development on carbonate islands. In:Budd DA, Saller AH, Harris PA (eds) Unconformities in carbonate strata – their recognition and the significance of associated porosity. American Association of Petroleum Geologists Memoir, Tulsa, OK 63:55–76

Nicols G (1999) Sedimentology and stratigraphy. Blackwell, Oxford 355

Patterson LJ, Sturchio NC, Kennedy BM, van Soet MC, Sultan M, Lu Z-T, Lehmann B, Purtschert R, El Alfy Z, El Kaliouby B, Dawood Y, Abdallah A (2005) Cosmogenic, radiogenic, and stable isotopic constraints on groundwater residence time in the Nubian Aquifer, Western Desert of Egypt. Geochem Geophys Geosyst 6:1–19

Peneva EL, Stanev EV, Stanychni SV, Salokhiddinov A, Stulina G (2004) The recent evolution of the Aral Sea level and water properties: analysis of satellite, guage and hydrometeorological data. J Mar Syst 47:11–24

Phillips FM, Castro MC (2003) Groundwater dating and residence time measurements. In: Drever JI (ed) Surface and ground water, weathering, and soils (Vol. 5, p. 451–497). In: Holland HD and Turekian, K. (eds) Treatise on geochemistry. Oxford University Press, Oxford

Pillans B (1983) Upper quaternary marine terrace chronology and deformation, South Taranaki, New Zealand. Geology 11:292–297

Plummer LN, Preston EC, Parkhurst DL (1991) NETPATH- An interactive code (NETPATH) for modeling net geochemical reactions along a flow path. US Geol Surv Water Res Invest Rep 91–4078, 94 pp

Ramirez-Herrera M-T, Urrutia-Fucugauchi J (1999) Morphotectonic zones along the coast of the Pacific continental margin, southern Mexico. Geomorphology 28:237–250

Rappleye HS (1948) Manual of leveling computation and adjustment, US Coast Geod Surv Spec Publ 240. US Dept of Commerce, Washington D.C. 94 pp

Reading HG (1996) Sedimentary environments: processes, facies and stratigraphy. Blackwell Science, Oxford

Rhoades R, Sinacori MM (1941) Patterns of groundwater flow and solution. J Geol 49:785–794

Richards DA, Smart PA, Edwards RL (1994) Maximum sea levels for the last glacial period from U-series ages of submerged speleothems. Nature 367:357–360

Rostami K, Peltier WR, Mangini A (2000) Quaternary marine terraces, sea-level changes and uplift history of Patagonia, Argentina: comparisons with predictions of the ICE-4G (VM2) model of the global process of glacial isostatic adjustment. Quat Sci Rev 19:1495–1525

Scicchitano G, Antonioli F, Castagnino Berlinghieri EF, Dutton A, Monaco C (2008) Submerged archaeological sites along the Ionian coast of southeastern Sicily (Italy) and implications for the Holocene relative sea-level change. Quat Res 70:26–39

Shalev E, Lazar A, Wollman S, Kington S, Yechieli Y, Gvirtzman H (2009) Freshwater–saltwater mixing zone in coastal aquifers: biased vs. reliable monitoring. Groundwater 47:49–56

Sneh (1979) Late Pleistocene fan deltas along the Dead Sea rift. J Sediment Petrol 49:541–552

Stute M, Schlosser P, Clark JF, Broecker, WS (1992) Paleotemperatures in the southwestern United States derived from noble gases in ground water. Science 256:1000–1002

Swinnerton AC (1932) Origin of limestone cavern. Geol Soc Am Bull 43:662–694

Taniguchi M, Iwakawa H (2001) Measurements of submarine groundwater discharge rates by a continuous heat-type automated seepage meter in Osaka Bay, Japan. J Groundwater Hydrol 43:271–277

Thrailkill J (1968) Chemical and hydrologic factors in the excavation of limestone caves. Geol Soc Am Bull 79:19–45
Ustun A, Demirel H (2006) Long-range geoid testing by GPS-leveling data in Turkey. J Surv Eng 132:15–23
Vanicek P, Krakiwsky EJ (1986) Geodesy: the concepts. North-Holland, Amsterdam
Voss CI (1984) SUTRA: finite-element simulation model for saturated–unsaturated, fluid-density-dependent groundwater flow with energy transport or chemically-reactive single species solute transport. US Geol Surv Water Resour Invest 84–4369:409
Waldmann N, Starinsky A, Stein M (2007) Primary carbonates and Ca-chloride brines as monitors of a paleo-hydrological regime in the Dead Sea basin. Quat Sci Rev 26:2219–2228
Williams PW (1982) Speleothems dates, quaternary terraces and uplift rates in New Zealand. Nature 298:257–260
Xu C, Liu J, Song C, Jiang W, Shi C (2000) GPS measurements of present-day uplift in the Southern Tibet. Earth Planets Space 52:735–739
Yechieli Y (2000) Fresh–saline water interface in the western Dead Sea area. Groundwater 38:615–623
Yechieli Y (2006) The response of the groundwater system to changes in the Dead Sea level. In: Enzel Y, Agnon A, Stein M (eds) New frontiers in Dead Sea paleoenvironmental research. Geological Society of America Special Paper, Boulder, CO 401:113–126
Yechieli Y, Abelson M, Bein A, Crouvi O, Shtivelman V (2006) Sinkhole "swarms" along the Dead Sea coast: reflection of disturbance of lake and adjacent groundwater systems. Geol Soc Am Bull 118:1075–1087
Yechieli Y, Kafri U, Goldman M, Voss CI (2001) Factors controlling the configuration of the fresh–saline water interface in the Dead Sea coastal aquifer: synthesis of TDEM surveys and numerical groundwater modeling. Hydrogeol J 9:367–377
Yechieli Y, Kafri U, Sivan O (2009a) The inter-relationship between coastal sub-aquifers and the Mediterranean Sea, deduced from radioactive isotopes analysis. Hydrogeol J 17:265–274
Yechieli Y, Kafri U, Wollman S, Shalev E, Lyakhovsky V (2009b) The effect of base level changes and geological structures on the location of the groundwater divide, as exhibited in the hydrological system between the Dead Sea and the Mediterranean Sea. J Hydrol 378:218–229
Yechieli Y, Ronen D, Berkovitz B, Dershovitz WS, Hadad A (1995) Aquifer characteristics derived from the interaction between water levels of a terminal lake (Dead Sea) and an adjacent aquifer. Water Resour Res 31:893–902

Chapter 5
Capturing of Groundwater Basins and Shifts of Divides

5.1 General

The boundaries of surface water drainage systems and watersheds are controlled by both base-level elevation and topography. Changes of the above might occur due to erosion, backward incision of the drainage systems and as a result capturing and shifts of the surface water divides.

When dealing with groundwater basins, additional factors might effect capturing, diversion of groundwater flow from one basin to another, and divide shifts. The response of the systems to those factors can be in some cases faster than in the case of surface water drainage basins. The main effecting factors involved in the above processes are climatic changes and the resultant changes of recharge, considerable tectonic changes of base-level elevation, karstification, groundwater sapping and anthropogenic groundwater exploitation.

5.2 Climate Changes

Groundwater systems as well as lake levels are sensitive to both global and local climate changes. Pluvial and humid climates result in increased recharge to the groundwater systems, and thus rising water tables and adjoining lake levels that are fed by both ground and surface water. Closed topographic depressions are transferred to closed basin lakes that start to act as local base-levels that attract convergent groundwater flow into them and as a result groundwater divides are formed in between those lakes (Urbano et al. 2004). Consequently, the entire regional groundwater flow system is subjected to partitioning into smaller cells (Urbano et al. 2000).

Arid climates, on the other hand, result in decreased surface flows, higher evapotranspiration, reduced recharge to groundwater and decline of both groundwater levels and lake levels, sometimes attaining a complete desiccation. These

U. Kafri and Y. Yechieli, *Groundwater Base Level Changes and Adjoining Hydrological Systems*,
DOI 10.1007/978-3-642-13944-4_5, © Springer-Verlag Berlin Heidelberg 2010

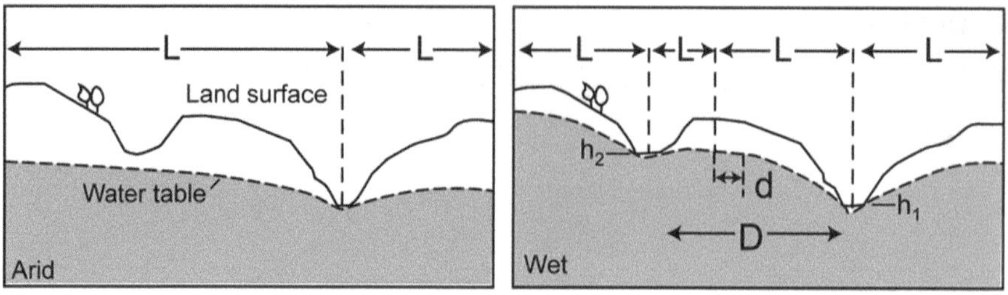

Fig. 5.1 Conceptual model indicating the changes of water table geometry between *wet* and *dry periods* [after Urbano et al. (2004)]. During the arid period, the h_2 base level is captured to the lower h_1 base level

lakes cease to act as local base-levels or discharge zones, the local groundwater divides disappear and groundwater flow is being diverted and captured to lower (lakes) base-levels (Urbano et al. 2004) (Fig. 5.1). The shift of the groundwater divides is related to the new consequences of recharge distribution, to the slope of the water table and the loss of water through evaporation from the lakes as demonstrated by simulations (Gosselin and Khisty 2001).

Smith et al. (1997) described the same phenomenon from Minnesota, USA where lakes in the northern Great Plains responded to climatic change as follows: Local (lake) base-levels, which were formed by rising water tables during humid periods, were subsequently, during Holocene droughts and water table decline, captured to lower base-levels.

Holzbecher (2001) described, from the north-east German Lake District, the dynamics of groundwater divides between the groundwater dominated lakes. Lake Stechlin and the neighboring lakes are closed basins with no surface flow. Both hydrological observations and modeling have shown that the system is sensitive to current climatic changes whereby the anthropogenic effect is negligible. Groundwater watersheds change and the groundwater divides between the lakes build-up, move or vanish as a consequence of changes from rainy years to dry ones.

Variations in aquifer recharge, due to climatic changes, are also able to modify the groundwater flow network so as to move position of groundwater divides, as also stated by Dragoni and Sukhija (2008).

5.3 Tectonically Induced Elevation Changes of Base-Levels

Tectonic displacements might result in formation of a new base-level close to an already existing one which will end up capturing part of the existing groundwater basin to newly formed one. This phenomenon is emphasized in cases of newly formed continental base-levels close to the marine base-level and often below sea level. Example of such a case is the Dead Sea Rift (DSR) base level (Kafri 1970;

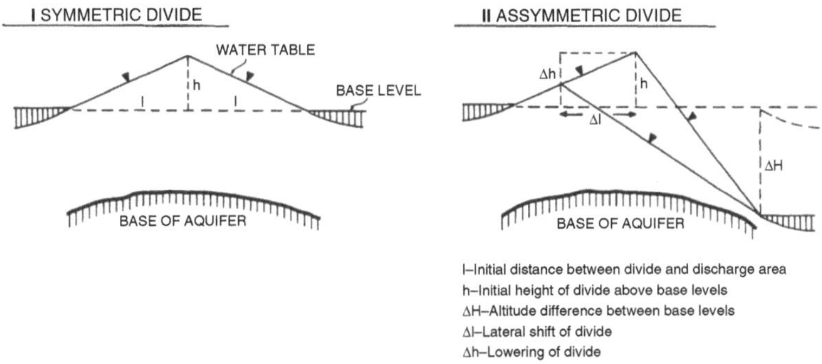

Fig. 5.2 Schematic model showing a divide shift and capturing of part of the groundwater basin due to subsidence of one base level [after Kafri and Arad (1978)]

Kafri and Arad 1978, 1979), as discussed in detail elsewhere in this book (see also Sect. 11.2).

The process consists basically of the following consecutive stages as schematically described in Fig. 5.2. In the first stage, a symmetric groundwater divide exists between two base levels of the same elevation. Following tectonic movement, which disconnects these base levels and a subsequent formation of a new subsiding continental base level, the location of the groundwater divide between these base levels is changed. The newly formed steep groundwater table toward the continental base level tends to moderate itself, thereby resulting in a shift of the divide toward the upper marine base level and capturing of part of the previous basin to the continental base level.

It should be emphasized that such a process of capturing, through a divide shift, can only take place in those cases where the base of the aquifer system between the base-levels is below the base-levels and the divide's location is not controlled by an elevated structure underneath. Indeed, simulations of past and future base-level elevation changes showed only minor lateral divide shifts in the structurally elevated regions (Yechieli et al. 2009a, b) (see also Sect. 11.2.6.4).

5.4 Karst Systems

Karstic aquifers contain dissolution-generated conduits that permit rapid transport of groundwater often across shallow divides. These groundwater basin divides are not firmly fixed and they might shift depending on the rate of discharge. As a result, phenomena of groundwater piracy routes into adjacent groundwater basins are common (White 2002). As an example, a study of a karst system in Brazil (Auler 1998) describes a capturing phenomenon of a groundwater system from one base-level to another one. The system is at present drained to the Velhas River base-level whereas in the past, based on cave passages distribution, it drained to a different depression that acted as a base-level.

5.5 Stream Piracy by Groundwater Sapping

Local groundwater divides often exist between two streams that are fed by the same groundwater system. As described in a model (Fig. 5.3) (Pederson 2001), a groundwater divide exists between adjacent streams of different elevations. Through a groundwater sapping process, which increases the aquifer porosity by subsurface weathering of emerging groundwater, the divide disappears and the upper stream area is being pirated in the subsurface to the lower stream. Pederson (2008) proposed the same process for the capturing of the old Colorado River system, associated with groundwater capturing to lower base-levels. The same mechanism of groundwater sapping through groundwater level decline was suggested to be responsible for deep canyons incision in the Andes of northern Chile (Hoke et al. 2004).

5.6 Groundwater Exploitation

Groundwater exploitation, and especially over-exploitation, can form deep cones of depression in the piezometric surface within a large groundwater basin, which may modify the original steady state flow regime. The drop of the piezometric surface can spread until it reaches the basin boundaries or divides and capture parts of a neighboring groundwater basin. Examples to the above are numerous. Eckstein et al. (2001) described a drop of both the groundwater table and the adjoining level of Crystal Lake in Ohio, which is located north-north-west of the continental groundwater divide between the St. Lawrence River basin and the Ohio-Mississippi River basin. According to their interpretation, following excessive groundwater

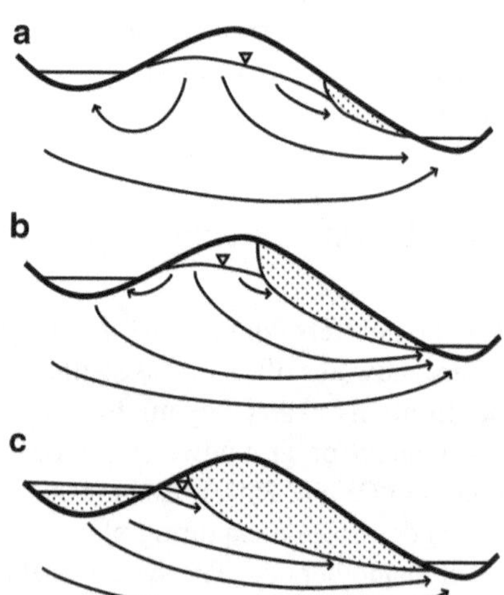

Fig. 5.3 Cross-section sequence (**a–c**) of pirating channel extension and eventual divide breaching by groundwater sapping [after Pederson (2001)]. *Arrows* show paths of groundwater flow

exploitation and water table decline, the groundwater divide shifted and part of the flow pattern was reversed, capturing groundwater to the Mississippi basin.

Similarly, Coon and Sheets (2006) described basically the same phenomenon on a regional scale. According to their report, the effect of heavy pumping in the Chicago and Milwaukee areas has caused a westward shift of the groundwater divide between the Great Lakes basin and the Mississippi River basin.

References

Auler AS (1998) Base-level changes inferred from cave paleoflow analysis in the Lagoa Santa karst, Brazil. J Cave Karst Stud 60:58–62

Coon WF, Sheets RA (2006) Estimate of groundwater in storage in the Great Lakes Basin, United States, 2006. US Geol Surv Sci Invest Rep 2006–5180:7

Dragoni W, Sukhija BS (2008) Climate change and groundwater: a short review. Geol Soc Lond Spec Publ 288:1–12

Eckstein Y, Matyjasik B, Matyjasik M (2001) Dislocation of the continental ground-water-divide induced by excessive pumping in Summit County, Ohio. Hydrol Sci Technol 17:119–128

Gosselin DC, Khisty MJ (2001) Simulating the influence of two shallow, flow-through lakes on a groundwater system: implications for groundwater mounds and hinge lines. Hydrogeol J 9:476–486

Hoke GD, Isacks BL, Jordan TE, Yu JS (2004) Groundwater-sapping origin for the giant quebradas of northern Chile. Geology 32:605–608

Holzbecher E (2001) The dynamics of subsurface water divides-watersheds of Lake Stechlin and neighbouring lakes. Hydrol Process 15:2297–2304

Kafri U (1970) Factors controlling the location of the groundwater divide in northern Israel. J Hydrol 11:22–29

Kafri U, Arad A (1978) Paleohydrology and migration of the ground-water divide in regions of tectonic instability in Israel. Geol Soc Am Bull 89:1723–1732

Kafri U, Arad A (1979) Current subsurface intrusion of Mediterranean seawater. A possible source of groundwater salinity in the Rift Valley system, Israel. J Hydrol 44:267–287

Pederson DT (2001) Stream piracy revisited: a groundwater sapping solution, GSA Today 11:4–10

Pederson JL (2008) The mystery of the pre-Grand Canyon Colorado River-results from the Muddy Creek formation. GSA Today 18:4–10

Smith AJ, Donovan JJ, Ito E, Engstrom DR (1997) Ground-water processes controlling a prairie lake's response to middle Holocene drought. Geology 25:391–394

Urbano LD, Person M, Hanor J (2000) Groundwater–lake interactions in semi-arid environments. J Chem Explor 69–70:423–427

Urbano LD, Person M, Kelts K, Hanor JS (2004) Transient groundwater impacts on the development of paleoclimatic lake records in semi-arid environments. Geofluids 4:187–196

White WB (2002) Karst hydrology: recent developments and open questions. Eng Geol 65:85–105

Yechieli Y, Kafri U, Wollman S, Shalev E, Lyakhovsky V (2009a) The effect of base level changes and geological structures on the location of the groundwater divide, as exhibited in the hydrological system between the Dead Sea and the Mediterranean Sea. Geol Surv Isr Rep GSI/10/2009, 22 pp

Yechieli Y, Kafri U, Wollman S, Shalev E, Lyakhovsky V (2009b) The effect of base level changes and geological structures on the location of the groundwater divide, as exhibited in the hydrological system between the Dead Sea and the Mediterranean Sea. J Hydrol 378:218–229

Chapter 6
Salinity, Salination and Freshening of the Different Base-Levels and Their Adjoining Groundwater Systems

6.1 General

Following the description of the main base-level types (Chap. 2), the salinity configurations and processes that prevail in these different base-levels and their adjoining groundwater systems, are discussed herein. In addition, the main mechanisms of salination and freshening that occur, both as a result of base-level changes and anthropogenic activity, are also described. The description basically deals with the two main types of base-levels, the global marine base-level and the continental endorheic base-levels. Regarding the latter, special attention is paid to endorheic base-levels that are considerable below sea level that are close to the sea (see also Chap. 11).

6.2 The Global Marine Base-Level

6.2.1 Base-Level Salinity

Normal ocean water salinity over most of the oceans' area exhibits a long term stability with a total TDS concentration around 3.5 g/l. Somewhat higher salinities of >3.8 g/l are observed in semi closed seas that are connected to the ocean, such as the Mediterranean Sea or the Red Sea.

6.2.2 Adjoining Groundwater Systems Salinity

The salinity of coastal aquifers was discussed in numerous studies from all over the world. The basic factors that control the salinity of the coastal aquifers are the prevailing sea level, the normal seawater salinity and the adjoining groundwater

U. Kafri and Y. Yechieli, *Groundwater Base Level Changes and Adjoining Hydrological Systems*,
DOI 10.1007/978-3-642-13944-4_6, © Springer-Verlag Berlin Heidelberg 2010

table levels or piezometric surfaces. Due to the salinity and density differences between fresh groundwater and normal seawater (100 mg/l, 1 gr/cm^3 and 3.5 g/l, 1.025 g/cm^3 respectively) the seawater encroaches the coastal aquifers by a density driven mechanism, in a steady state condition. An interface occurs between the encroaching seawater body and the overlying fresh water body (Fig. 6.1). The configuration and depth of the interface follow the Ghyben–Herzberg approximation given by the equations:

$$h_f = \alpha h_s \tag{6.1}$$

$$\alpha = (\rho_s - \rho_f)/\rho_f \tag{6.2}$$

where h_f is the elevation of the water table above sea level, h_s is the depth of the interface below sea level and ρ_f and ρ_s being the densities of the fresh and saline water respectively.

Salination of coastal aquifers can take place as a result of natural processes of sea level rise due to climate changes (Sherif and Singh 1999) or to tectonic and isostatic changes (see also Chap. 3). Sea level rise results in an inland seawater encroachment into the coastal aquifers, whereas a sea level decline causes the opposite process. In addition, salination of coastal aquifers often takes place in numerous coastal aquifers in the world as a result of groundwater exploitation and mostly through over-exploitation. The process of depletion of groundwater levels of coastal aquifers, due to over-exploitation, results in a shallowing of the interface followed by a farther inland seawater penetration or upconing of saline waters from underneath the formed water table cones of depression.

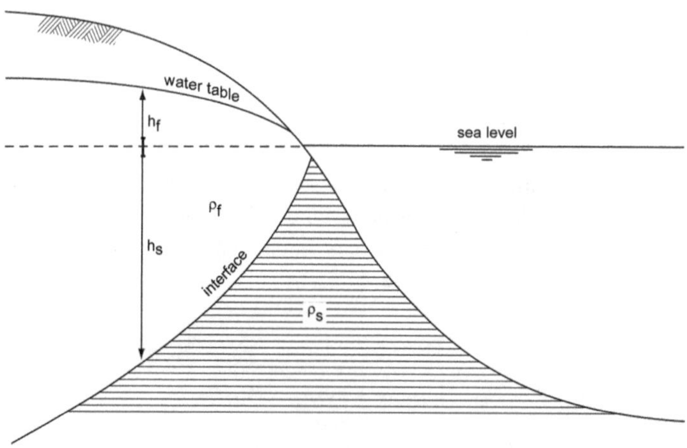

Fig. 6.1 The configuration of the interface between fresh groundwater and intruding saline water. h_f and h_s denote the elevation of the water table above sea level and the depth of the interface below sea level, respectively. ρ_f and ρ_s are the densities of the fresh and saline water respectively

Coastal sabkhas or lagoons can also be considered as part of the marine base-level system. They are partly or temporary connected to the sea and thus differ, regarding their salinity, from normal seawater. Their salinity is higher than that of seawater due to their combination of restricted connection with the sea and often high evaporation related to aridity. In coastal features, such as in Abu Dhabi (Fig. 6.2) (Patterson and Kinsman 1981; Sanford and Wood 2001), there is a hydraulic connection between the sea and the sabkha. This connection is usually via an inlet or surface flooding of seawater either at strong storms or at times of high sea level. The subsurface hydraulic connection between the sabkha and the sea is quite complicated since it involves three different water bodies, namely seawater, the brine of the sabkha and relatively fresh groundwater. The interface between the brine and the seawater is expected to incline toward the sea bottom, unlike the situation in regular marine coastal aquifer (Yechieli and Wood 2002). A variation of such a system exists where a shore ridge or bar prevents direct flooding of seawater and permits only subsurface connection (e.g., in Kuwait, Robinson and Gunatilaka 1991; Sinai, Gavish et al. 1985) as shown in Fig. 6.3. An example of water and solute balance in a typical coastal sabkha is given by Yechieli and Wood (2002).

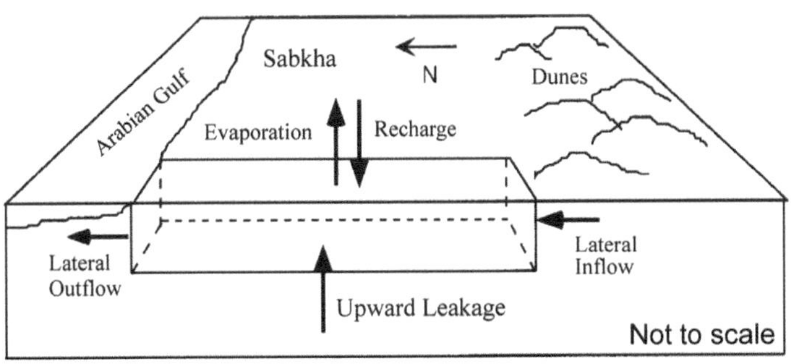

Fig. 6.2 Hydrological configuration of a large coastal sabkha in Abu Dhabi [after Sanford and Wood (2001)]

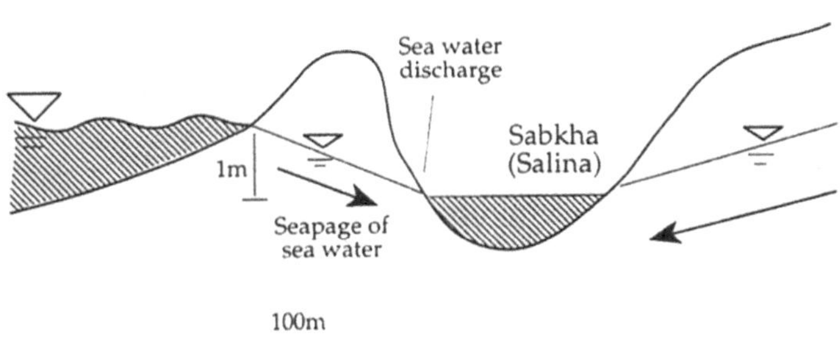

Fig. 6.3 Hydrological configuration of a coastal sabkha in Sinai, Egypt [after Gavish et al. (1985)]

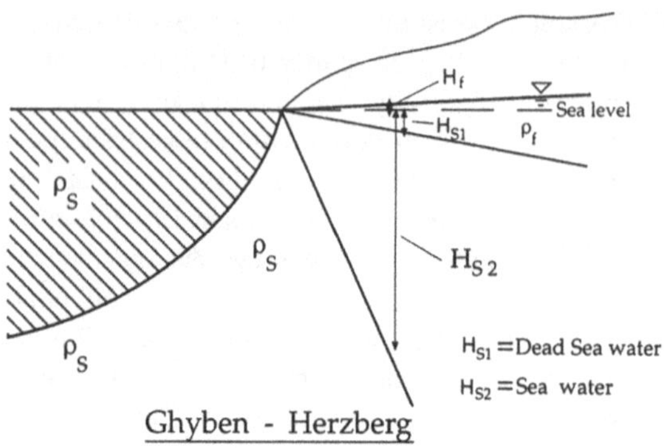

Fig. 6.4 The slope of different fresh–saline water interfaces according to the different densities of the saline water bodies [after Yechieli (1993)]

The salinity of those coastal sabkhas is, thus, affected by the following:

(a) Sea level rise is supposed to increase the surface flooding, even several kilometers inland, in cases of a flat topography, or to increased subsurface encroachment of the sabkha by seawater.
(b) The arid environment is responsible for the relatively minor input of freshwater from the coastal margin.
(c) Due to the formation of a concentrated brine within the sabkha, an interface is formed also between the brine body within the sabkha and the mainland coastal fresh water body. In this case, due to the higher density difference between both water bodies, the interface is shallower as compared to a normal seawater case (Fig. 6.4) (Yechieli 1993; Yechieli and Wood 2002) and, thus in turn, its toe penetrates more inland the coastal aquifer.

6.3 Continental Endorheic Base-Levels Distant from the Sea

6.3.1 Base-Level Salinity

This category includes terminal closed base-levels, which are usually subjected to convergent of both surface and groundwater flows. In most cases they are located in arid regions where evaporation exceeds fresh water input from all sources, namely rainfall, surface water and groundwater inflows (Langbein 1961; Hardie et al. 1978). This combination, by itself, is sufficient to result in a saline environment. The salinity of lakes is generally controlled by the volume of fresh water input, which is determined by the amount of precipitation in the entire drainage basin which is controlled by the global or local climate.

The typical saline environments that prevail in the discussed base-levels are hypersaline lakes and inland Sabkhas (in Arabic), Salinas, Salars and Playas (in Spanish), and Pans, Salt pans or Salt flats (in English). A summary of the terminology of these saline environments is given by Rosen (1994). Saline lakes are common in hot climates but also exist in many locations typical of a cold climate, such as in Canada (Last and Slezak 1988) and Antarctica (Lyons et al. 1998).

Due to the negative water balance, salinity in saline lakes gradually increases to brine concentrations that results in sequential deposition of evaporites, such as gypsum and anhydrite, halite and post-halite salts. Salination also takes place in the sabkhas or playas which act as discharge zones to the convergent groundwater flow whereby evaporation occurs from the seasonally fluctuating shallow groundwater level where the capillary fringe is close to the surface. An additional important source of salinity is the process of leaching or dissolution of pre-existing evaporites, whether by fresher water (i.e., Risacher and Fritz 2000; Portugal et al. 2005) or by ascending thermal water (i.e., Faure et al. 2002). Salination of closed base-levels was also attributed to ascending thermal, mostly Ca-Cl brines in the Andean Altiplano (Lowenstein and Risacher 2009), in the Death Valley (Larsen et al. 2001; Lowenstein and Risacher 2009), and in the Qaidam Basin, China (Lowenstein and Risacher 2009) base-levels.

An additional mechanism of salination was suggested by an upward convective seawater flow, related to hydrothermal anomalies, associated with young volcanic activity, as described from the Salton Trough by Barragan et al. (2001).

Examples to all the above mentioned processes are described in more detail in Chaps. 9–11.

6.3.2 *Adjoining Groundwater Systems Salinity*

The hydrogeological setup, regarding the salinity of the closed endorheic base-levels and their adjoining groundwater systems was summarized, among others, schematically by Fan et al. (1997) and Holzbecher (2005) (Fig. 6.5). The model

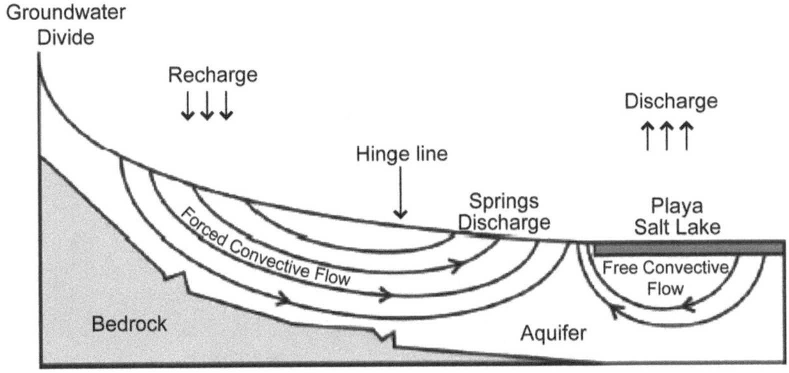

Fig. 6.5 The conceptual model of groundwater circulation in a desert closed basin [modified after Fan et al. (1997) and Holzbecher (2005)]

describes a convergent convective fresh water flow from the margins of the basin to the center of the base-level, development of brines and a resultant mechanism of lateral density driven flow in the opposite direction, to the aquifer. A similar mechanism was described for the DSR base-level by Stanislavsky and Gvirtzman (1999). An additional process which has to do with a hypersaline lake base-level overlying a fresher groundwater system underneath, is described from the Mono Lake, California (i.e., Rogers and Dreiss 1995a, b) and in Australia (Simmons and Narayan 1997). The proposed mechanism is of a downward advection flow of the brine from the lake to the groundwater system due to the instability created by the density differences. This process increases the salinity in groundwater below and around the lake. An opposite direction of advective flow from the groundwater system to the overlying saline lake appears as springs and manifested as tufa deposits (see also Sect. 9.6).

Base-level changes are expected to affect the salinity of the adjoining groundwater systems as follows: A drop of a saline lake level, due to gradual desiccation, is being followed by increasing salinity of waters occupying this lake, that serves as a base-level, and consequently expected to cause a parallel drop of the adjoining groundwater levels. This process, in turn, may cause a gradual freshening of the upper portion of the dewatered groundwater system, due to a basin-ward retreat of the fresh/saline water interface (Fig. 6.6). An increase in the salinity of groundwater nearby the saline lake may still occur in the lower part of the aquifer (below the interface), even in the case of decreasing lake levels, since re-circulating of lake water continues when the system reaches a quasi steady state situation (Kiro et al. 2008). A base-level rise, on the other hand, due to more humid conditions, is expected to cause the opposite process which includes the raising of the adjoining groundwater levels, and the reduction of the salinity of the base-level. The expected effect on groundwater salinity is rather complicated. The farther inland shift of the

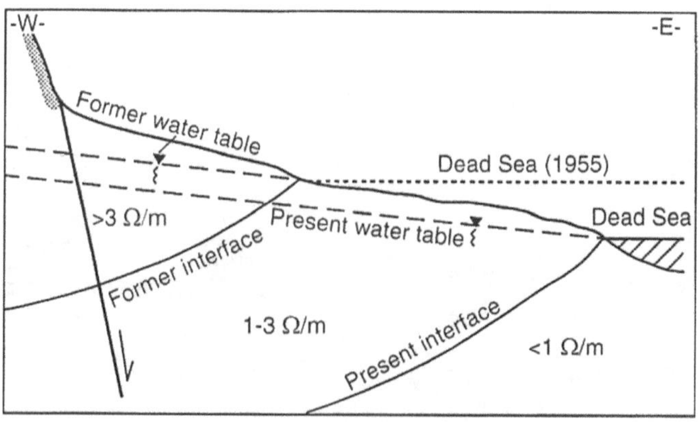

Fig. 6.6 A schematic basin-ward shift of the fresh–saline water interface following a decline of the Dead Sea base level, exhibited by resistivity (TDEM) data [after Kafri et al. (1997)]

fresh/saline water interface will increase the salinity of groundwater at various parts of the aquifer. On the other hand, the lower salinity of the lake and thus its inland encroachment of a lower salinity will result in a reduced water salinity below the interface. Only in the extreme case of complete change of the base-level to a fresh water lake, a complete flushing and freshening of the groundwater system is expected.

6.4 Flow-Through, Intermediate, Base-Levels

6.4.1 Base-Level Salinity

This type of base-level is not a terminal endorheic base-level by definition. Despite the fact that this type of base-level attracts convergent groundwater flow, it still drains constantly or temporarily surface or subsurface flows down-gradient. Such base-levels are often aligned along regional endorheic base-levels, partly interconnected and ultimately draining to the lowermost terminal, completely closed and saline base-level. As such, most flow-through lakes are not subjected to the typical salinity concentration process of the terminal base-levels, thus remaining relatively fresh. These connected base-levels, via a shallow sill or a shallow groundwater system, can be diverted temporarily to a terminal endorheic base-level following a considerable drop of their level and turn to a saline base-level and vice versa.

In some cases, despite the fact that a lake acts as a flow-through base-level, it is in fact saline or hypersaline due to extreme aridity and the low ratio between the inflow and the outflow from the lake. The Caspian Sea, serves as an example to such a setup (see also Sect. 10.5). Examples to flow-through base-levels are known, among others, from the Chad Basin, Andean Altiplano, the Main Ethiopian Rift, the Caspian Sea and the Dead Sea Rift systems (see also Sects. 9.2, 9.4, 9.6, 10.5, 11.2, respectively).

6.4.2 Adjoining Groundwater Systems Salinity

The salinity of the groundwater systems related to this type of base-level is basically controlled by the same factors that control other types of base-levels described before. Here, also, salinity of the adjoining groundwater system depends on the base-level (lake) salinity, at a given time. In those cases only of a saline base-level, salination or freshening of the system as well as the shift of the fresh/saline water interface will be assumingly controlled by the changes of the lake level, as described before.

6.5 Salinity of Endorheic Base-Levels Below Sea Level, Close to the Sea

Special attention is paid herein to the category of continental base-levels below sea level and close to the sea regarding their salinity and salination mechanism. On top of all other processes, described before, which are responsible to salination of continental endorheic base-levels, an additional one is proposed herein. This process is current subsurface seawater encroachment from the upper marine base-level to the lower endorheic one. A detailed discussion of this process and the supporting evidences to its existence are detailed in Chap. 11.

6.6 Freshening of Coastal Aquifers

Salination and freshening of coastal aquifers, whether related to the marine or continental lacustrine base-levels, as mentioned before, are often related to base-level changes. A coastal aquifer, which was at a certain time encroached by sea water or a lacustrine brine, could have been at a later stage subjected to flushing and freshening processes. These could happen following a subsequent drop of its marine or lacustrine base-level and/or a rise of the land, resulting in the relative drop of the adjoining fresh water table and the saline/fresh water interface (Fig. 6.7).

As a consequence of the above, the upper portion of the aquifer, which was previously below the water table and/or the interface, is being diverted to an unsaturated zone since most of its water has been drained to the base-level. During the drop of the groundwater level, the saturated saline portion of the aquifer is partly flushed and replaced by fresh aquifer water (e.g., the Dead Sea coastal aquifer, Sect. 11.2.6).

Freshening of coastal aquifers can also be the result of freshening of their adjoined lake's water due to climate change. At a more humid climate, more fresh water (rain and floods) drain to the lake reducing its salinity and thus reducing the salinity of the saline encroachment into the aquifer.

Except of the above physical processes, chemical ones are also involved which including diagenetic processes in the aquifer which affect the chemical composition of the groundwater. As an example, dedolomitization of a Cretaceous marine carbonate sequence exposed to fresh water above the fresh/saline water transition zone, was described by Magaritz and Kafri (1981). Similarly, stable isotope changes of Cretaceous carbonate concretions were attributed to same processes, related to sea level fluctuations (Coniglio et al. 2000). Diagenetic processes of groundwater through base exchange, related to the freshening process were also described in detail by Lambrakis and Kallergis (2001) and Lambrakis (2006).

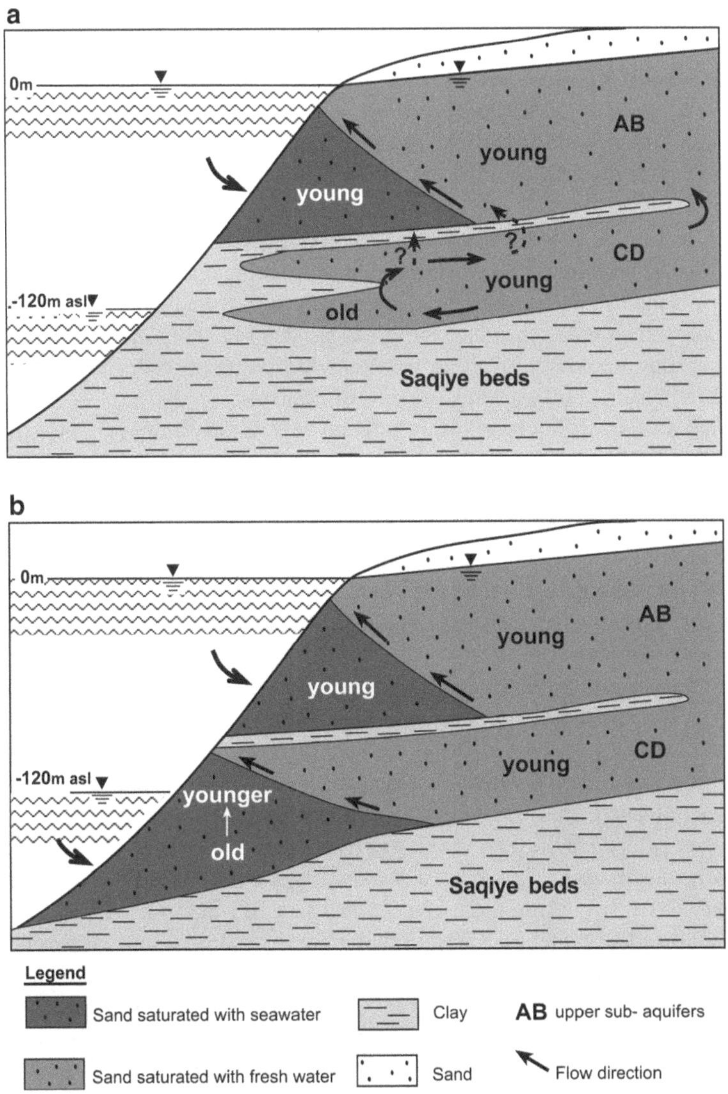

Fig. 6.7 The effect of sea level rise on saline groundwater ages in the vicinity of the Mediterranean Sea [after Yechieli et al. (2009)]. Note the upward younger ages of intruded seawater into the Mediterranean coastal aquifer of Israel subsequent to the last glacial maximum. Two options are given: (**a**) the lower sub-aquifer is connected to the sea; (**b**) the lower sub-aquifer is not connected to the sea

The dynamics and duration of such freshening processes was tackled, among others, by the following studies:

Voss and Andersson (1993) described the rate of flushing of seawater in the Baltic shield since the Holocene. According to them, the Baltic Shield was isostatically depressed by the weight of the ice cap, some 10 ka BP, to over 200 m below sea level and thus encroached by seawater. Flushing of the seawater to the newly formed marine base-level started following the subsequent rise of the Baltic Shield and the retreat of the base-level coastlines due to the snow melt. The incomplete Holocene flushing of sea water since then is attributed to both slow fluid movement through fractures in the bedrock and long travel distances to the base-level.

Groen et al. (2000) and Kooi and Groen (2003) used groundwater flow modeling to study the freshening of the pore waters in submarine aquifers, tens of kilometers offshore Suriname. They found that groundwater movement of the primary flow system was too slow to reproduce the observed deep flushing of the aquifers. They concluded that the wedge of the offshore meteoric groundwater is not part of an active flow system but was formed due to exposure of the aquifers to flushing during the Wisconsin sea level drop in the last glacial maximum (LGM) when sea level was at 120 m bsl.

Kafri et al. (1997) and Yechieli et al. (2001) discussed the duration of flushing of brines from the coastal aquifers bordering the Dead Sea base-level following its drop since 14 ka BP and especially during the last decades. Their conclusion was that the "exposed", unsaturated sequence of the former aquifers is almost completely drained and flushed and that a complete flushing seems to usually require only some several decades.

In some cases, freshening can occur in newly coastal areas formed due to base-level drop. An example was described from the Dead Sea shore, where a new unsaturated zone has been formed in the last 50 years (Yechieli 1993).

The time span during which the Mediterranean coastal aquifer of Israel was completely flushed from previously intruded seawater was also discussed by Yechieli et al. (2009). The working hypothesis was that the aquifer was completely flushed out of its hosted seawater when the Mediterranean base-level was some 120 m bsl during the LGM, some 18 ka BP. That would require that the presently hosted seawater in the aquifer is related to the gradual encroachment of seawater to the aquifer due to the gradual rise of the sea level since then and are, thus, younger than 18 ka BP. Indeed, ^{14}C dating of intruded seawater samples yielded an age sequence whereby preliminary younger ages were found upward in the sequence, as expected (Fig. 6.7) (Yechieli et al. 2009). Thus, it seems that a period of several tens of thousands years was more than sufficient for a complete flushing of the aquifer.

Manzano et al. (2001) described paleo-fresh groundwater that occupy the coastal aquifers of Spain, Mallorca and Gran Canaria. This groundwater yielded ages between 6 and 15 ka BP, younger than the age of the Late Pleistocene (LGM) base-level low-stand some 18 ka BP. One can arrive at the conclusion, as done before, that waters that pre-dated that age had sufficient time to be completely flushed from the system.

Lambrakis and Kallergis (2001) estimated, based on the duration of the chemical processes, that freshening time of sea water intruded granular aquifers in Greece is between 8,000 and 10,000 years. For carbonate karstic aquifers they estimated that only a 15 years freshening time duration is required. The difference assumingly stems from the fact that the carbonate aquifers have a higher transmissivity, as compared to the granular ones, and thus are more rapidly flushed.

References

Barragan RRM, Birkle P, Portugal ME, Arellano GVM, Alvarez R (2001) Geochemical survey of medium temperature geothermal resources from the Baja California Peninsula and Sonora, Mexico. J Volcanol Geotherm Res 110:101–119

Coniglio M, Myrow P, White T (2000) Stable carbon and oxygen isotope evidence of Cretaceous sea-level fluctuations recorded in septarian concretions from Pueblo, Colorado, USA. J Sediment Res 70:700–714

Fan Y, Duffy CJ, Oliver Jr DS (1997) Density-driven groundwater flow in closed desert basins: field investigations and numerical experiments. J Hydrol 196:139–184

Faure H, Walter RC, Grant DR (2002) The coastal oasis: ice age springs on emerged continental shelves. Glob Planet Change 33:47–56

Gavish E, Krumbein WE, Friedman GM (1985) Geomorphology, mineralogy and groundwater geochemistry as factors of the hydrodynamic system of the Gavish Sabkha. In: Friedman GM, Krumbein WE (eds) Hypersaline ecosystems, Springer New York, 186–217

Groen J, Velstra J, Meesters GCA (2000) Salinization processes in paleowaters in coastal sediments of Suriname: evidence from $d^{37}Cl$ analysis and diffusion modeling. J Hydrol 234:1–20

Hardie LA, Smoot JP, Eugster HP (1978) Saline lakes and their deposits: a sedimentological approach. In: Matter A, Tucker ME (eds), Modern and ancient lake sediments. International Association of Sedimentary Special Publication, London 2:7–41

Holzbecher E (2005) Groundwater flow pattern in the vicinity of a salt lake. Hydrobiology 532:233–242

Kafri U, Goldman M, Lang B (1997) Detection of subsurface brines, freshwater bodies and the interface configuration in-between by the time domain electromagnetic method in the Dead Sea Rift, Israel. Environ Geol 31:42–49

Kiro Y, Yechieli Y, Lyakhovsky V, Shalev E, Starinsky A (2008) Time response of the water table and saltwater transition zone to a base level drop. Water Resour Res 44:W12442, DOI:10.1029/2007WR006752

Kooi H, Groen J (2003) Geological processes and the management of groundwater resources in coastal areas. Geol Mijnb 82:31–40

Lambrakis N (2006) Multicomponent heterovalent chromatography in aquifers. Modelling salinization and freshening phenomena in field conditions. J Hydrol 323:230–243

Lambrakis N, Kallergis G (2001) Reaction of subsurface coastal aquifers to climate and land use changes in Greece: modeling of groundwater refreshening patterns under natural recharge conditions. J Hydrol 245:19–31

Langbein WS (1961) Salinity and hydrology of closed lakes. U S Geol Surv Prof Pap 412:1–20

Larsen D, Swihart GH, Xiao Y (2001) Hydrochemistry and isotope composition of springs in the Tecopa basin, southeastern California, USA. Chem Geol 179:17–35

Last WM, Slezak LA (1988) The salt lakes of western Canada: a paleolimnological overview. Hydrobiology 158:301–316

Lowenstein TK, Risacher F (2009) Closed basin brine evolution and the influence of Ca-Cl inflow waters: death valley and Bristol dry lake California, Qaidam basin, China, and Salar de Atacama, Chile. Aquat Geochem 15:71–94

Lyons WB, Welch KA, Neumann K, Toxey JK, McArthur R, Williams C (1998) Geochemical linkages among glaciers, streams and lakes within the Taylor Valley, Antarctica. In: Priscu JC

(ed) Ecosystem dynamics in a polar desert. The McMurdo Dry Valleys, Antarctica. Antarctic research series. American Geophysical Union, Washington, DC 72:77–92
Magaritz M, Kafri U (1981) Stable isotope and Sr^2/Ca^2 evidences of diagenetic dedolomitization in a schizohaline environment: Cenomanian of northern Israel. Sediment Geol 28:29–41
Manzano M, Custodio E, Loosli H, Cabrera MC, Riera X, Custodio J (2001) Paleowater in coastal aquifers of Spain. Geol Soc Lond Spec Publ 189:107–138
Patterson RJ, Kinsman DJJ (1981) Hydrologic framework of a sabkha along the Arabian Gulf. Am Assoc Petrol Geol Bull 66:28–43
Portugal E, Izquierdo G, Truesdell A, Alvarez J (2005) The geochemistry and isotope hydrology of the Southern Mexicali Valley in the area of the Cerro Prieto, Baja California (Mexico) geothermal field. J Hydrol 313:132–148
Risacher F, Fritz B (2000) Bromine geochemistry of the salar de Uyuni and deeper salt crusts, Central Altiplano, Bolivia. Chem Geol 167:373–392
Robinson BW, Gunatilaka A (1991) Stable isotope studies and hydrological regime in southern Kuwait, Arabian Gulf. Sediment Geol 73:141–159
Rogers DB, Dreiss SJ (1995) Saline groundwater in Mono Basin, California 1. Distribution. Water Resour Res 31:3131–3150
Rogers DB, Dreiss SJ (1995) Saline groundwater in Mono Basin, California 2. Long-term control of lake salinity by groundwater. Water Resour Res 31:3151–3169
Rosen MR (1994) The importance of groundwater in playas: a review of playa classification and the sedimentology and hydrology of playas. Geol Soc Am Spec Pap 289:1–18
Sanford WE, Wood WW (2001) Hydrology of the coastal Sabkhas of Abu Dhabi, United Arab Emirates. Hydrogeol J 9:358–366
Sherif MM, Singh VP (1999) Effect of climate change on sea water intrusion in coastal aquifers. Hydrol Proc 13:1277–1287
Simmons CT, Narayan KA (1997) Mixed convection processes below a saline disposal basin. J Hydrol 194:263–285
Stanislavsky E, Gvirtzman H (1999) Basin-scale migration of continental-rift brines: paleohydrological modeling of the Dead Sea basin. Geology 27:791–794
Voss CI, Andersson J (1993) Regional flow in the Baltic Shield during Holocene coastal regression. Ground Water 31:989–1006
Yechieli Y (1993) The effects of water level changes in closed lakes (Dead Sea) on the surrounding groundwater and country rocks. PhD Thesis, Weizmann Institute of Science, Rehovot, Israel
Yechieli Y, Wood WW (2002) Hydrogeologic processes in saline systems: playas, sabkhas, and saline lakes. Earth Sci Rev 58:343–365
Yechieli Y, Kafri U, Goldman M, Voss CI (2001) Factors controlling the configuration of the fresh-saline water interface in the Dead Sea coastal aquifer: synthesis of TDEM surveys and numerical groundwater modeling. Hydrogeol J 9:367–377
Yechieli Y, Kafri U, Sivan O (2009) The inter-relationship between coastal sub-aquifers and the Mediterranean Sea, deduced from radioactive isotopes analysis. Hydrogeol J 17:265–274

Chapter 7
Paleo, Current and Future Marine Base-Levels

7.1 General

The oceans and their connected seas served, all along their geological history, as the ultimate or terminal base-levels to both surface and groundwater flow from the continents. Internal endorheic closed depressions, seas and lakes, also existed on the continents serving as base-levels to convergent flows into them. In the absence of direct hydrological data regarding the older geological time spans, only an approximate rough paleohydrological reconstruction can be obtained based on available geological and sedimentological data.

Paleo-hydrological information, which enables a more accurate palaeo-hydrological reconstruction and delineation of paleo base-levels, becomes more abundant as somebody deals with the time windows of younger periods, which are already within the hydrological memory, namely since the Neogene and especially since the last glacial period. Most of those younger paleo base-levels still serve as base-levels.

The continental paleo base-levels that continue to serve currently as base-levels are not discussed in this chapter but are addressed and described in detail in the description of the specific current continental base-levels (Chaps. 9–11). Examples of known marine paleo-base- levels are described below.

7.2 The Neogene Marine Mediterranean Base-Level

The most famous and studied example of marine Neogene base-levels is the Parathetys-Mediterranean Sea Basin. The history of the Mediterranean Basin, and its connection and isolation from the Atlantic Ocean since the Neogene, was discussed in numerous studies. Among these are: Clauzon et al. (1996), Martin et al. (2001), Mocochain et al. (2006), Popov et al. (2006), Rouchy and Caruso (2006), Nunn and Harris (2007) and Govers (2009). Some of the authors differ as to

U. Kafri and Y. Yechieli, *Groundwater Base Level Changes and Adjoining Hydrological Systems*,
DOI 10.1007/978-3-642-13944-4_7, © Springer-Verlag Berlin Heidelberg 2010

the exact timing of the changes that prevailed in the basin, yet its history can be summarized in short as follows:

Until the end of Tortonian-early Messinian (Late Neogene), the Mediterranean Basin was connected to the Atlantic Ocean via two sea ways, namely the Rifian Corridor of northern Morocco and the Betic Strait in southern Spain (Fig. 7.1) (Martin et al. 2001; Nunn and Harris 2007).

The subsequent partial or complete closure and isolation of the Mediterranean Sea from the Atlantic Ocean is evidenced by two phases of thick evaporite sequences in the basin. The isolation was initiated by the Altic orogenesis, or alternatively by isostatic uplift (Govers 2009), which closed the Betic Corridor around 7.6–7.5 Ma and caused the shallowing of the Rifian Corridor, creating restricted estuarine circulation and a subsequent drop of sea level at 5.8 Ma. The complete closure and isolation of the Mediterranean Sea, named the Messinian Salinity Crisis (MSC), between 5.6 and 5.32 Ma resulted in the choking of, and the partial or complete desiccation of the Mediterranean Sea. The sea level dropped to a depth exceeding 1,500 m, and according to canyons incision even to over 2,800 m (Nunn and Harris 2007). The surficial isolation from the Atlantic was complete, but Nunn and Harris (2007) claimed that partial subsurface seawater seepage, from the Atlantic Ocean to the declining and desiccating Mediterranean Sea, still occurred through the barrier that existed in between. A somewhat similar model, related to current subsurface seawater intrusion to continental base-levels below sea level and close to the sea, was previously suggested by Kafri and Arad (1979) and Kafri (1984) (see also Sect. 11.2). This same model is proposed herein to similar current continental base-levels in other locations around the world (Chap. 11).

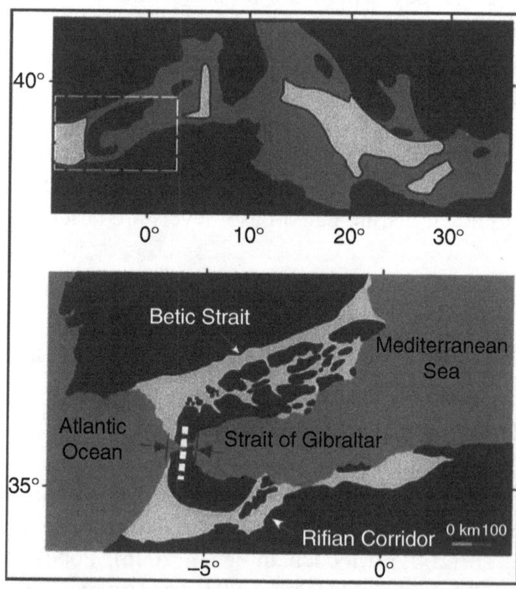

Fig. 7.1 The connection between the Atlantic Ocean and the Messinian Mediterranean Sea through the Betic Strait and the Rifian Corridor [after Nunn and Harris (2007)]

At a later stage, re-flooding of the Mediterranean basin by Atlantic water, through the Straits of Gibraltar, had started in the Pliocene (Popov et al. 2006) or some 5.33 Ma ago (Nunn and Harris 2007).

The relationship between karst systems and the different Mediterranean base levels since the Neogene was discussed, among others by Audra et al. (2004) and Mocochain et al. (2006). The changes of the karst levels reflect the changes of the paleo groundwater regime, related to the base level changes. During the time span of the MSC, the Mediterranean basin served as a deep closed endorheic basin and a base-level to deeply incised canyons as well as to convergent groundwater flow from the continents. The drop of the paleo-groundwater tables, following the Messinian decline of the base-level is indicated, for example, in the Rhone Valley System, and other drainage systems around the Mediterranean Sea by deep paleo-karst systems that led to that Messinian deep base-level at around 1,500 m bsl (Fig. 7.2). A shallower karst system level was formed later on as a result of the subsequent rise the Pliocene base-level and the adjoining paleo groundwater levels to a level around 80 m asl. The newly formed karst systems that led to this level were also fed by the previous deeper systems that were already chocked as a result of the rising base level. The Pliocene karst systems were later abandoned following the later drop of the base-level in Quaternary times and especially in the LGM and younger systems were formed leading to the newly formed base level at 120 m bsl. These were subsequently chocked following the rise of the Mediterranean base level to its present level. The above described paleohydrological history of the Mediterranean basin and its adjoined karst system is exhibited schematically in Fig. 7.2.

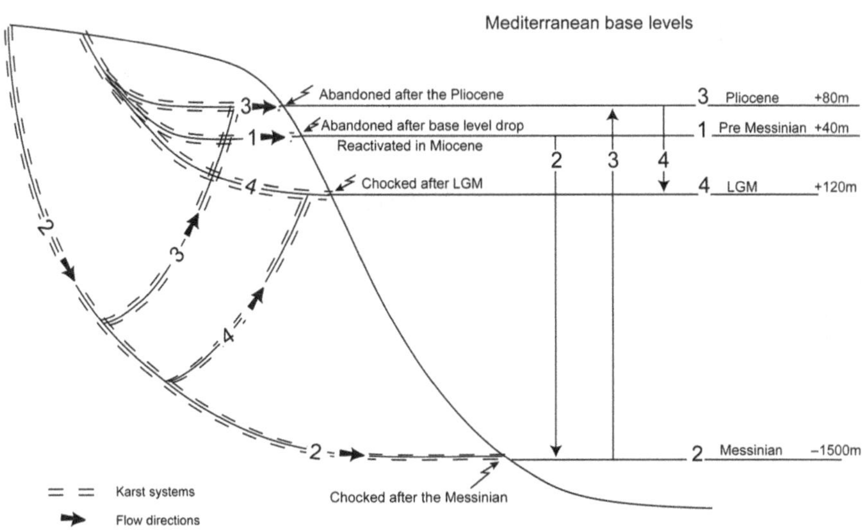

Fig. 7.2 The response of the paleo karstic and groundwater systems to Mediterranean base level changes since the Neogene [based on Audra et al. (2004) and Mocochain et al. (2006)]

The Mediterranean Sea, the Caspian Sea and the Black Sea were parts of and are remnants of the huge Tethys Ocean. The Caspian and the Black seas were connected to the ocean through the Mediterranean Sea until they were disconnected and became landlocked endorheic basins in the Late Miocene some 7.2 Ma ago (Zubakov 2001). The Caspian basin was subsequently subjected to sea level fluctuations and temporal connections to the Mediterranean basin through the Black Sea (Dumont 1998) (see also Sect. 11.5).

7.3 The Black Sea Neogene to Holocene Base-Levels

The Black Sea Basin (Fig. 7.3) is a sea arm that is connected to the Mediterranean Sea via the narrow straits of the Bosporus and the Dardanelles and the Sea of Marmara. As such, it serves at present as part of the Mediterranean Sea base-level, whereas in the past it alternated between a sea arm and a closed or semi-closed continental lake and base-level. The Black Sea is fed by the big Danube, Dniepr, Don and Dniestr rivers. Groundwater flow into the Black Sea is evident from the coastal aquifers in the periphery of the sea. It is manifested, as an example, by submarine groundwater discharge (SGD) from the offshore of Crimea utilizing the fault systems (Borisenko 2001) and by seawater encroachment to coastal aquifers in northeastern Bulgaria (Pulido-Bosch et al. 1997).

As mentioned before, the Black Sea, as well as the Caspian Sea, were interconnected and connected to the Mediterranean Sea already in late Neogene times,

Fig. 7.3 Map of the Black Sea and its surrounding

between 5.3 and 3.3 Ma (Zubakov 2001). Throughout the last 2 Ma the ancient Black Sea was basically a closed fresh water lake that was invaded by Mediterranean seawater during global sea level highstands when those exceeded the elevation of the Bosporus and Dardanelles sills (Ryan et al. 2003). When the Mediterranean Sea level dropped to below the sills' level, during cold climates, the Black Sea expanded and spilled to the Marmara Sea or farther to the Mediterranean Sea through the straits. On the other hand, during warm periods, the Black Sea was subjected to shrinkage (Ryan et al. 2003).

Throughout the last 900 ka, the Black Sea level fluctuated between roughly the present sea level and 150 m bsl (Fig. 7.4) (Svitoch et al. 2000) or 200 m bsl as claimed by Kislov and Toropov (2007). The discussed fluctuations were reconstructed using seashore terraces, alluvial fans and dated core samples. Black Sea level changes were basically controlled by the interrelationship between climatic changes, global ocean and Mediterranean Sea levels and by the amount of east European rivers inflow to the Black Sea (Svitoch et al. 2000; Winguth et al. 2000; Aksu et al. 2002; Kislov and Toropov 2007). A Black Sea level curve for the last 900 ka (Winguth et al. 2000) exhibits regressive phases of the sea during most of this time span with a minimum of 151 m bsl in the LGM and only shorter transgressive periods in between. This configuration somewhat differs from the one given by Svitoch et al. (2000) that describe longer transgressive periods during the same time span. In any case, it is clear that most of the time in the regressive (lowstand) phases the Black Sea level was below that of the present one. Moreover, in those periods when either the Mediterranean or the Black Sea levels dropped below the Bosporus-Dardanelles sill levels (see below), both basins were disconnected and it is logical to conclude that the Black Sea acted as a continental endorheic closed basin.

Special attention was given, by several authors, to the Mediterranean–Black Sea interrelationship during the LGM and the Holocene. As mentioned before, the Black Sea is at present connected to the Mediterranean Sea by the straits of the Bosporus and Dardanelles which constitute underneath a sill between the two basins. The elevation of the Dardanelles sill is approximately 85 m bsl and that of the Bosporus gorge is some 100 m bsl, filled with sediments to an elevation of

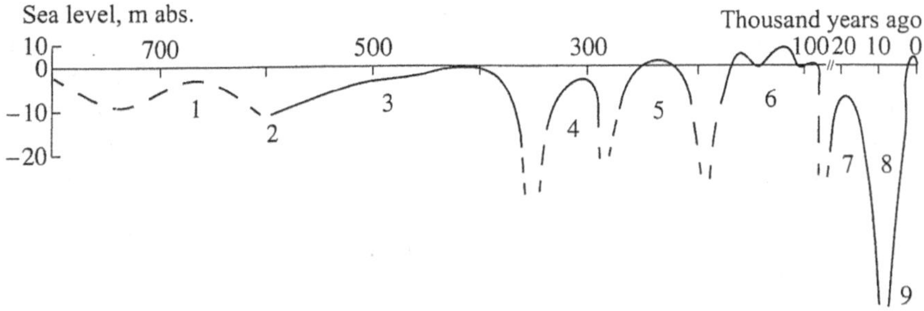

Fig. 7.4 Variations in the level of the Black Sea in the Pleistocene [after Svitoch et al. (2000)]

32 m bsl (Major et al. 2002). During the LGM the Mediterranean level was 120 m bsl and then subsequently started to rise until attaining the present level. A debate exists among the authors as to the exact time when the rising sea level exceeded the elevation of the sills enabling a Mediterranean Sea overflow to the Black Sea Basin, based on sedimentary, geochemical, isotopic and faunistic indicators. Ryan et al. (1997, 2003) proposed a catastrophic flooding of the Black Sea by the rising Mediterranean Sea level around 8.4 and 7.1 ka, based on a sill elevation of 85 m bsl. It is interesting to note, in this regard, that in addition to ancient sunken ships and submerged shorelines, a neolithic settlement, some 8,000 years old, was discovered at the sea bottom at a depth of approximately 100 m (Coleman and Ballard 2007). Major et al. (2002), however, raised the possibility that, considering a lower sill elevation, the Mediterranean transgression to the Black Sea could have occurred even earlier, around 12.8 ka. The catastrophic flooding process and the timing of the transgression were opposed and argued by Bahr et al. (2006) who proposed Mediterranean water inflow to the Black Sea between 9 and 8 ka, and by Balabanov (2007) who proposed a gradual sea level rise, rather than a catastrophic flooding, during that time span.

It is, therefore, evident that the Black Sea was subjected to sea level fluctuations, acting in certain periods as a continental closed base level. It is logical to assume that the peripheral paleo groundwater systems followed the base-level changes, namely declining groundwater levels in lowstand periods accompanied by flushing of the aquifers, and rising groundwater levels in highstand periods, accompanied by sea or lake water encroachment into the aquifers.

The area which connects the Black Sea to the Mediterranean Sea, namely the Sea of Marmara and the Bosporus and Dardanelles straits, is structurally a plate boundary. The area is dissected by major fault systems which include strike slip and normal faults that form pull apart basins, such as the Sea of Marmara (i.e., Wong et al. 1995; Armijo et al. 2002). The fault system is still active as evidenced by the seismicity of the region (i.e., Okay et al. 2000).

Another possibility that should be taken into account, regarding Mediterranean Sea connection with the Black Sea, is subsurface Mediterranean seawater encroachment to the Black Sea, proposed herein. This might have happened in cases when both basins were below the sills' elevation and thus not surficially connected, and the Mediterranean Sea level was higher than the Black Sea level. Such subsurface hydraulic connection could exist through subsurface conductive formations that are found in between. The dense fault systems could have served in the past, in addition, as preferential conduit systems to subsurface flows. A comparison of the global (or Mediterranean) sea level curve (Miller et al. 2005) for the last one million years with the Black Sea level of the same period (Winguth et al. 2000) indicates that such a configuration could have existed during a few time spans. It is, thus, suggested herein that subsurface Mediterranean water inflow to the Black Sea could have taken place in the past in a somewhat similar manner to that proposed for the current cases described in this book (see also Chap. 11).

7.4 Quaternary and Holocene Marine Base-Levels

It was reported in various studies that continental shelves occupy underneath them aquifers that contain fresh to brackish waters. Such aquifers are known to extend, in places, to a distance of more than a hundred kilometers offshore the coastlines (Fig. 7.5).

The combination that favors presently a development of an extensive fresh-water wedge in the offshore requires a high groundwater head at the coastline and the submarine aquifer to be a thick, semi-confined, and a high permeability aquifer (Kooi and Groen 2001). According to them, the above factors by themselves do not always prevail and the occurrence of submarine fresh waters is too far offshore to be solely explained by a current active flow system. Therefore, they concluded that these waters are Pleistocene paleowaters that were recharged and were flowing in the aquifers when the marine base-levels were lower than at present.

As described in numerous studies (i.e., Fairbanks 1989) and mentioned before, the global sea level dropped at the peak of the LGM, some 18–20 ka BP, to 120 m

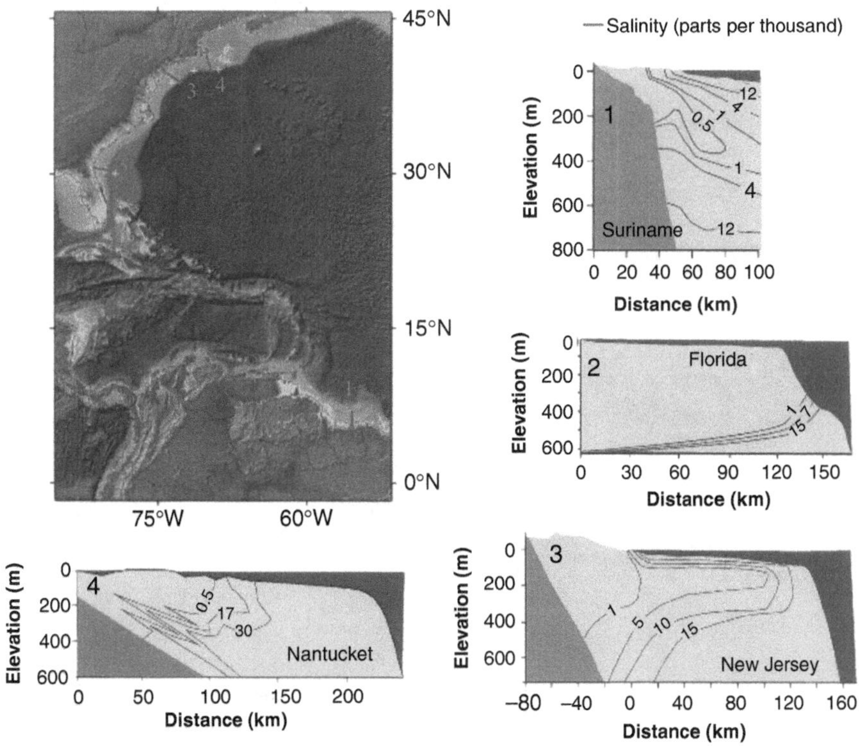

Fig. 7.5 Fresh groundwater tongues offshore Suriname and the US Atlantic coast as exhibited by groundwater salinity (ppt) concentrations [after Cohen et al. (2009)]

below the present sea level. This level became the global marine base-level to the adjoining on-land aquifers. The drastic drop of the sea level from the considerably higher previous interglacial highstand resulted in a steepening of the water table gradients, increasing piezometric heads and enhanced groundwater flow toward the Holocene systems (Faure et al. 2002).

Arad (1983) described the paleo-water system of the continental shelf of Guyana which extends tens of kilometers offshore draining to the 130 m bsl, base-level. The subsequent rise of the sea level to the present day one reversed the process and fresh groundwater was displaced by intruding seawater land wise.

The same paleo-system was also described from offshore Suriname as well as from offshore the Atlantic coast of the US (Fig. 7.5) and Indonesia (Groen et al. 2000; Kooi and Groen 2003; Cohen et al. 2009). By using groundwater flow modeling, they concluded that the present flow system is too slow to reproduce the observed deep flushing of the aquifers and thus the offshore fresh groundwater wedge was formed through exposure of the aquifers to flushing during the Wisconsin sea level drop during the LGM. Subsequently, due to the sea level rise, the process was reversed pushing the fresh–saline water interface landward. Moreover, Kooi and Groen (2003) showed that the deep coastal and offshore confined aquifers are over-pressured due to the load of the overlying Holocene sediments and compaction, resulting in a landward flow direction.

Person et al. (2003) described the fresh–saline water interface of the Atlantic continental shelf off New England which is far out of equilibrium with modern sea

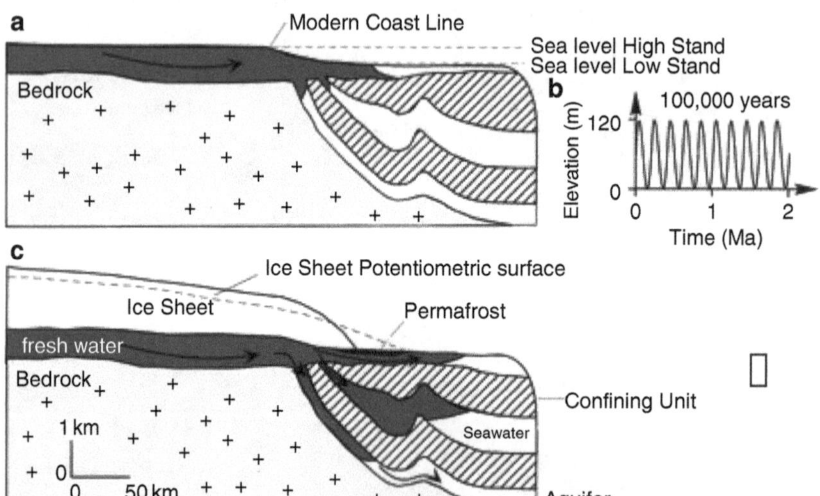

Fig. 7.6 Conceptual models explaining the incursion of fresh groundwater far out the Atlantic continental shelf of New England, USA [after Person et al. (2003)]. (**a**) Seaward incursion of fresh water is caused by repeated sea level lowstands during the Pleistocene; (**b**) assumed sea level changes in the Pleistocene; (**c**) Seaward fresh water incursion is enhanced by meltwater from the overlying ice sheet

level conditions. The existing interface configuration, according to them, requires a combination of both a Pleistocene low-stand base-level and in addition a subglacial recharge from the Laurentide Ice Sheet (Fig. 7.6).

Deep paleowaters from the offshore of the North Sea, the Danish coast, the English Channel as well as the coasts of France, Spain and Portugal were also described by Edmunds et al. (2001).

Similar to the above, the drainage of the vast Nubian Sandstone aquifer, which covers the Sahara and north Africa, to the Holocene Mediterranean base-level was described, among others, by Faure et al. (2002).

Yechieli et al. (2009) have shown from the Mediterranean coastal aquifer of Israel that a complete flushing of the aquifer probably occurred during the LGM low-stand base-level, prior to the subsequent rise of sea level. This is implied by the findings of preliminary dated waters within the aquifer, whether fresh or seawater, which are younger than the 18 ka BP, LGM age. Similar observations were reported from coastal paleowaters from Spain, Mallorca and Gran Canaria (Manzano et al. 2001).

It can be seen that paleo-marine base-levels, such as the LGM and the Holocene ones, played an important role in the regime of the adjoining groundwater systems in regard to groundwater flow, salination and flushing of coastal aquifers (see also Chap. 6).

7.5 Current and Future Marine Base-Levels

The oceans and their connected inter-continental seas, such as the Mediterranean Sea or the Red Sea, occupy some 71% of the total surface area of the globe. They serve as the ultimate base-level and discharge zone to both surface water and groundwater flows.

Groundwater flow to the sea takes place through regional and coastal aquifers that surround the oceans and the seas at the margins of the continents. They drain directly to the sea at the vicinity of the shoreline, to coastal lagoons or sabkhas or as SGD somewhat offshore (i.e., Taniguchi et al. 2002).

The levels of ancient marine base-levels were subjected to considerable changes during the geological history. Sea levels have changed as a result of alterations in climatic conditions and atmospheric circulation during the geological history.

Several theoretical studies examined the effect of sea level changes on the fresh–saline interface in a confined aquifer using sharp interface models (e.g., Essaid 1990; Harrar et al. 2001; Meisler et al. 1985) and density-dependent flow models (Ataie-Ashtiani et al. 1999; Kooi et al. 2000; Chen and Hsu 2004). Some of these studies claim that the response of the interface to sea level changes is rather slow and it is presently still in equilibrium with Pleistocene conditions (e.g., Meisler et al. 1985).

The climate is currently subjected to a global warming process, which is largely influenced by human activity leading to the global warming, as claimed by several studies and opposed by others. In any case, as a result of the global warming process, whether natural or anthropogenic, polar icebergs and on land glaciers are melting and whereby raising the sea level. A sea level rise of ~0.5 m by the year 2100 was suggested by Warrick et al. (1996). This gradual rise of the global marine base level is expected to affect the adjoining groundwater systems.

Several studies dealt with the effect of the sea level rise on groundwater flow and sea water intrusions (i.e., Leatherman 1984; Navoy 1991; Sherif and Singh 1999; Oude Essink 1996, 2001; Shrivastava 1998; Kooi et al. 2000; Bobba

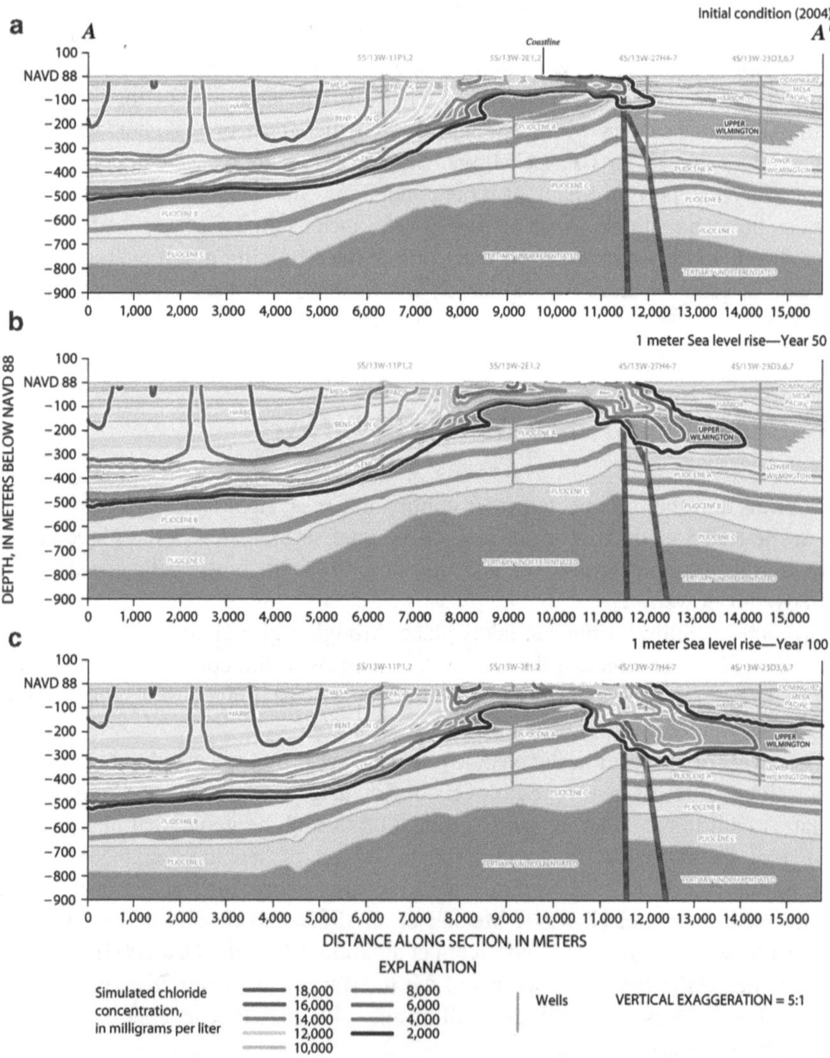

Fig. 7.7 The effect of future sea level rise on seawater intrusion into an aquifer in California, USA [after Nishikawa et al. (2009)] Contours are of simulated chloride concentrations assuming a 1-m sea-level rise for model scenarios of: (**a**) 2004 (initial condition), (**b**) year 50, and (**c**) year 100

2002; Feinson 2001; Tyagi 2005; Chachadi and Lobo-Ferreira 2005; Werner and Simmons 2009). According to these studies, a sea level rise will cause a salination of coastal aquifers. A 50 cm sea level rise is expected to cause an additional inland seawater intrusion of 9 km in the Nile Delta and 0.4 km in the Bengal Bay (Sherif and Singh 1999) as well as an increase of groundwater salinity in the Netherlands near the shore (Oude Essink 1996). A time lag in the salination is expected where the permeability is low, and in areas far from the shore (Oude Essink 1996). A time lag is also expected if the sea level rise is fast (Kooi et al. 2000). A vertical salination above the fresh water may occur, due to seawater flooding, when the permeability is high and the sea level rise is fast (Kooi et al. 2000). Deep coastal aquifers with low hydraulic gradients were found to be more vulnerable to sea water encroachment following sea level rise (Sherif and Singh 1999).

The effect of sea level rise on seawater intrusion and salination of a coastal aquifer depends significantly on the slope of the coastal topography. A steep coast was found to permit only a short distance inland seawater encroachment, following the expected sea level rise, as simulated for the Israeli coastal aquifer (Yechieli et al. 2010). The specific stratigraphic sequence of the coastal aquifer determines the actual extent of the seawater encroachment (Nishikawa et al. 2009) (Fig. 7.7).

References

Aksu AE, Hiscott RN, Kaminski MA, Mudie PJ, Gillespie H, Abrajano T, Yasar D (2002) Last glacial-Holocene paleoceanography of the Black Sea and Marmara Sea: stable isotopic, foraminiferal and coccolith evidence. Mar Geol 190:119–149

Arad A (1983) A summary of the artesian coastal basin of Guyana. J Hydrol 63:299–313

Armijo R, Meyer B, Navarro S, King G, Barka A (2002) Asymmetric slip partitioning in the Sea of Marmara pull-apart: a clue to propagation processes of the North Anatolian Fault? Terra Nova 14:80–86

Ataie-Ashtiani B, Volker RE, Lockington DA (1999) Tidal effects on sea water intrusion in unconfined aquifers. J Hydrol 216:17–31

Audra P, Mocochain L, Camus H, Gilli E, Clauzon G, Bigot JY (2004) The effect of the Messinian Deep Stage on karst development around the Mediterranean Sea. Examples from southern France. Acta Geodin 17:27–38

Bahr A, Arz HW, Lamy F, Wefer G (2006) Late glacial to Holocene paleoenvironmental evolution of the Black Sea, reconstructed with stable oxygen isotope records obtained on ostracod shales. Earth Planet Sci Lett 241:863–875

Balabanov IP (2007) Holocene sea-level changes of the Black Sea. In: V. Yanko-Hombach et al. (eds). The Black Sea flood question. Springer, Berlin. 711–730

Bobba G (2002) Numerical modelling of salt-water intrusion due to human activities and sea-level changes in the Godavari Delta, India. Hydrol Sci J 47: S67–S80

Borisenko LS (2001) Hydrogeological conditions of submarine groundwater discharge in the Crimea. Water Resour 28:15–21

Chachadi AG, Lobo-Ferreira JP (2005) Assessing aquifer vulnerability to sea-water intrusion using GALDIT method: Part 2 – GALDIT Indicator Descriptions. IAHS and LNEC, Proceedings of the 4th The Fourth Inter Celtic Colloquium on Hydrology and Management of Water Resources. Universidade do Minho, Guimarães, Portugal

Chen BF, Hsu SM (2004), Numerical study of tidal effects on seawater intrusion in confined and unconfined aquifers by time-independent finite-difference method, J. Water Port Coastal Ocean Eng., 130, 191–206, DOI:10.1061/(ASCE)0733-950X(2004)130:4 (191)

Clauzon G, Suc J-P, Gautier F, Berger A and Loutre M-F (1996) Alternate interpretation of the Messinian salinity crisis: controversy resolved? Geology 24:363–366

Cohen D, Person M, Wang P, Gable CW, Hutchinson D, Marksamer A, Dugan B, Kooi H, Groen K, Lizarralde D, Evans RL, Day-Lewis FD, Lane Jr JW (2009) Origin and extent of fresh paleowaters on the Atlantic continental shelf, USA. Ground Water 48:143–158

Coleman DF, Ballard RD (2007) Submerged paleoshorelines in the southern and western Black Sea-implications for inudated prehistoric archaeological sites. In: V. Yanko-Hombach et al. (eds). The Black Sea flood question. Springer, Berlin 671–696

Dumont HJ (1998) The Caspian lake: History, biota, structure, and function. Limnol Oceanogr 43:44–52

Edmunds WM, Hinsby K, Marlin C, Condesso de Melo MT, Manzano M, Vaikmae R, Travi Y (2001) Evolution of groundwater systems at the European coastline. Geol Soc Lond Spec Publ 189:289–311

Essaid HI (1990) A multilayered sharp interface model of coupled freshwater and saltwater flow in coastal systems: model development and application. Water Resour Res 26:1431–1454

Feinson LS (2001) The effects of sea level rise on coastal aquifers. MS Thesis. University of Nevada, Reno

Fairbanks RG (1989) A 17,000-year glacio-eustatic sea level record: influence of glacial melting rates on the younger Dryas event and deep-ocean circulation. Nature 342:637–642

Faure H, Walter RC, Grant DR (2002) The coastal oasis: ice age springs on emerged continental shelves. Glob Planet Change 33:47–56

Govers R (2009) Choking the Mediterranean to dehydration: the Messinian salinity crisis. Geology 37:167–170

Groen J, Velstra J, Meesters GCA (2000) Salinization processes in paleowaters in coastal sediments of Suriname: evidence from δ^{37}Cl analysis and diffusion modeling. J Hydrol 234:1–20

Harrar WG,Williams AT, Barker JA, Camp MV (2001) Modelling scenarios for the emplacement of palaeowaters on aquifer systems. In: Edmunds WM, Milne CJ (eds) Palaeowaters in coastal Europe: evolution of groundwater since the Late Pleistocene. Geological Society of London Special Publications, London 189:213–229

Kafri U (1984) Current subsurface seawater intrusion to base levels below sea level. Environ Geol Water Sci 6:223–227

Kafri U, Arad A (1979) Current subsurface intrusion of Mediterranean seawater. A possible source of groundwater salinity in the Rift Valley system, Israel. J Hydrol 44:267–287

Kislov A, Toropov P (2007) East European rivers runoff and Black Sea and Caspian sea level changes as simulated within the paleoclimate modeling intercomparison project. Quat Int 167–168:40–48

Kooi H, Groen J, Leijnse A (2000) Modes of seawater intrusion during transgressions. Water Resour Res 36(12):3581–3589

Kooi H, Groen J (2001) Offshore continuation of coastal groundwater systems; predictions using sharp-interface approximations and variable-density flow modeling. J Hydrol 246:19–35

Kooi H, Groen J (2003) Geological processes and the management of groundwater resources in coastal areas. Geol Mijnb 82:31–40

Leatherman SP (1984) Coastal geomorphic response to sea level rise: Galveston Bay, Texas. In: Barth MC, Titus JG (eds), Greenhouse effects and sea level rise: a challenge for this generation. Van Nostrand Reinhold Co, New York 151–178

Major C, Ryan W, Lericolais G, Hajdas I (2002) Constraints on Black Sea outflow to the Sea of Marmara during the last glacial–interglacial transition. Mar Geol 190:19–34

Manzano M, Custodio E, Loosli H, Cabrera MC, Riera X, Custodio J (2001) Paleowater in coastal aquifers of Spain. Geol Soc Lond Spec Publ 189:107–138

Martin JM, Braga JC and Betzler C (2001) The Messinian Guadalhorce corridor: the last northern, Atlantic-Mediterranean gateway. Terra Nova 13:418–424

Meisler H.P, Leahy P, Knobel LL (1985) Effect of eustatic sea level changes on saltwater–freshwater relations in the northern Atlantic coastal plain, US Geol Surv Water Supp Pap 2255:28

Miller KG, Kominz MA, Browning JV, Wright JD, Mountain GS, Katz ME, Sugarman PJ, Cramer BS, Christe-Blick N, Pekar SF (2005) The Phanerozoic record of global sea-level change. Science 310:1293–1298

Mocochain L, Clauzon G, Bigot J-Y, Brunet P (2006) Geodynamic evolution of the peri-Mediterranean karst during the Messinian and the Pliocene: evidence from the Ardeche and Rhone Valley systems canyons, southern France. Sediment Geol 188–189:219–233

Navoy AS (1991) Aquifer–estuary interaction and vulnerability of ground-water supplies to sea-level-rise driven saltwater intrusion. PhD Thesis. State College, Pennsylvania State University 225

Nishikawa T, Siade AJ, Reichard EG, Ponti DJ, Canales AG, Johnson TA (2009) Stratigraphic controls on seawater intrusion and implications for groundwater management, Dominguez Gap area of Los Angeles, California, USA. Hydrogeology 17:1699–1725

Nunn JA and Harris NB (2007) Subsurface seepage of seawater across a barrier: a source of water and salt to peripheral salt basins. Geol Soc Am Bull 119:1201–1217

Oude Essink GHP (1996) Impact of sea level rise on groundwater flow regimes, a sensitivity analysis for the Netherlands. PhD Delft University of Technology, p 411

Oude Essink GHP (2001) Saltwater intrusion in a three dimensional groundwater system in the Netherland: a numerical study. Transp Porous Media 43:137–158

Okay AI, Kaslilar-Ozcan A, Imren C, Boztepe-Guney A, Demirbag E, Cuscu I (2000) Active faults and evolving strike-slip basins in the Marmara Sea, northwest Turkey: a multichannel seismic reflection study. Tectonophysics 321:189–218

Person M, Ugan B, Swenson JB, Urbano L, Stott C, Taylor J, Willett M (2003) Pleistocene hydrogeology of the Atlantic continental shelf, New England. Geol Soc Am Bull 115: 1324–1343

Popov SV, Shcherba IG, Llyina LB, Nevesskaya LA, Paramonova NP, Khondkarian SO, Magyar I (2006) Late Miocene to Pliocene palaeogeography of the Paratethys and its relation to the Mediterranean. Palaeogeogr Palaeoclimatol Palaeoecol 238:91–106

Pulido-Bosch A, Machkova M, Lopez-Chicano M, Calvache ML, Dimitrov D, Calaforra JM, Velikov B, Pentchev P (1997) Hydrogeology of the upper aquifer, Dobrich Region, Northeastern Bulgaria. Hydrogeol J 5:75–85

Rouchy JM and Caruso A (2006) The Messinian salinity crisis in the Mediterranean basin: a reassessment of the data and an integrated scenario. Sediment Geol 188–189:35–67

Ryan WBF, Pitman III WC, Major CO, Shimkus K, Moskalenko V, Jones GA, Dimitrov P, Gorur N, Sakinc M, Yuce H (1997) An abrupt drowning of the Black Sea shelf. Mar Geol 138:119–126

Ryan WBF, Major CO, Lericolais G, Goldstein SL (2003) Catastrophic flooding of the Black Sea. Annu Rev Earth Planet Sci 31:525–554

Sherif M, Singh VP (1999) Effect of climate change on sea water intrusion in coastal aquifers. Hydrol Processes 13:1277–1287

Shrivastava GS (1998) Impacts of sea level rise on seawater intrusion into coastal aquifer. J Hydrol Eng 3:74–78

Svitoch AA, Selivanov AO, Yanina TA (2000) Paleohydrology of the Black Sea Pleistocene basins. Water Resour 27:594–603

Taniguchi M, Burnett WC, Cable JE, Turner JV (2002) Investigation of submarine groundwater discharge. Hydrol Processes 16:2115–2129

Tyagi AK (2005) Impact of sea level rise on saltwater intrusion in coastal aquifers. In: Raymond Walton PE (ed) Proceedings of the 2005 World Water and Environmental Resources Congress, Anchorage, Alaska

Warrick RA, Oerlemans J, Woodworth PL, Meier MF, Le Provost C (1996) Changes in sea level. In: Houghton JT, Meira Filho LG, Callander BA (eds) Climate changes 1995: the science of

climate, Contribution of Working Group I to the second assessment Report of the Intergovernmental Panel of Climate Changes. Cambridge University Press, Cambridge 359–405
Werner AD, Simmons CT (2009) Impact of sea-level rise on sea water intrusion in coastal aquifers. Groundwater 47:197–204
Winguth C, Wong HK, Panin N, Dino C, Georgescu P, Ungureanu G, Krugliakov VV, Podshuveit V (2000) Upper Quaternary water level history and sedimentation in the northwestern Black Sea. Mar Geol 167:127–146
Wong HK, Luedmann T, Ulug A, Gorur N (1995) The Sea of Marmara: a plate boundary sea in an escape tectonic regime. Tectonophysics 244:231–250
Yechieli Y, Kafri U, Sivan O (2009) The inter-relationship between coastal sub-aquifers and the Mediterranean Sea, deduced from radioactive isotopes analysis. Hydrogeol J 17:265–274
Yechieli Y, Shalev E, Wollman S, Kiro Y, Kafri U (2010) The response of the Mediterranean and Dead Sea coastal aquifers to sea level variations. Geol Surv Isr Rep GSI/11/2010, 22 pp
Zubakov VA (2001) History and causes of variations in the Caspian Sea level; the Miopliocene, 7.1–1.95 million years ago. Water Resour 28:249–256

Chapter 8
Characteristics of Current Continental Endorheic Base-Levels

Endorheic continental basins, also termed as terminal or closed basins, are internal drainage systems. The term is usually related to surface water drainage systems, yet endorheic basins often also serve as groundwater base-levels and discharge zones. The basins are typified by the fact that there is no outflow from the basin to the sea by surface rivers and the rain water over the basin leaves the system naturally only by evaporation or seepage. The bottom or the lower surface area of such basins is often occupied by salt lakes or salt pans.

Paleo-closed endorheic basins are assumed also to serve as base-levels to convergent surface as well as of groundwater flow. This is despite the fact that usually it is difficult or impossible to reconstruct flow paths or paleo-water tables due to the lack of information as compared to present day situation. Nevertheless, present day situation can assist in shedding light by comparison to paleo phenomena.

Several watersheds on earth constitute of a series of closed basins or endorheic lakes which are dominated or controlled by their adjoining groundwater systems, since the lakes and the adjoining groundwater levels are inter-connected. This is often evidenced by close correspondence between the lake level and adjoining groundwater table fluctuations (i.e., Almendinger 1990; Person et al. 2007).

The location of the groundwater dominated lakes is determined by the aerial intersection of the ground surface topography and the water table as shown in Fig. 5.1 (Urbano et al. 2004). When a lake is formed through a water table rise to the surface of a topographic closed depression, the lake's surface area becomes a discharge zone (Urbano et al. 2000). The closed basin lakes act as a large diameter pumping well as a result of the lake's water loss by evaporation. In such a setup, the seepage into the lake is concentrated along the lake margins (Almendinger 1990). This type of convergent groundwater flow into a lake discharge zone was also described as a centripetal groundwater flow pattern in a groundwater dependent ecosystem from southern Spain by Rodriguez-Rodriguez et al. (2008).

When a large basin or watershed contains a series of closed basin lakes, a partitioning of the regional flow system is evidenced, being subdivided into smaller cells with local groundwater divides in between them (Urbano et al. 2000). In such a large basin, following a rise of the groundwater table, more and more topographic

U. Kafri and Y. Yechieli, *Groundwater Base Level Changes and Adjoining Hydrological Systems*,
DOI 10.1007/978-3-642-13944-4_8, © Springer-Verlag Berlin Heidelberg 2010

depressions are intersected by the water table to form lakes that act as discharge zones. New groundwater divides are formed close to the mid point between the newly formed lakes (Fig. 5.1) (Urbano et al. 2004). Gosselin and Khisty (2001) have shown, for a setup similar to the above, by simulating the influence of shallow lakes on groundwater systems, that the location of the groundwater divides, mounds and hinge lines, is a consequence of the relationship between recharge, water table slope and the lake's water loss by evaporation.

Examples of such endorheic lake systems are described from the Murray Basin, Australia (Urbano et al. 2000, 2004), from Nebraska, USA (Gosselin and Khisty 2001), from southern Spain (Rodriguez-Rodriguez et al. 2008) and from Minnesota, USA (Smith et al. 1997).

Similar to lake systems, groundwater divides also exist between stream valleys which are enlarged by groundwater sapping and fed by groundwater. The locations of the divides are controlled by the elevation difference of the discussed valleys (Pederson 2001).

Endorheic basins are abundant all over the world. They exist in different dimensions, part of them close to the sea and part of them distant from the sea, located in the centers of the continents. Some of them are topographically above the present sea level whereas others are considerably below it. Within the latter category, the basins that are close to the sea are discussed in detail in the book, owing to the fact that their configuration plays a significant role in the processes and mechanisms of groundwater salination (see also Chap. 11).

The different current continental base-levels are reviewed in this book in regard to the combination of elements or prevailing conditions which typifies their specific configurations. The elements that were considered (Table 8.1) are as follows:

1. Distance from the sea, whereby base-levels in a distance that does not exceed several tens of kilometers are regarded as close to the sea and those exceeding this distance are defined as distant from the sea.
2. Base level elevation in relation to the present sea level, whether above or below it.
3. The main structural configuration of the base-level whether a rift, a trough or an intermountain basin.
4. Prevailing young or current tectonic deformation such as displacement or subsidence.
5. Prevailing young or current seismicity.
6. Prevailing young or current volcanism.
7. Prevailing young or current hydrothermal activity.
8. The extension of the base-level is manifested as an embayment or a gulf on its intersection with the sea shore.
9. The existence of the current base-level already as a paleo base-level.
10. The existence of the current base-level already as a paleo-sea arm.
11. An existing shallow groundwater divide between the continental and marine base-levels.
12. Prevailing arid climate.
13. The base-level occupies saline lakes, brines or salt pans.

Table 8.1 Characteristics of current marine and continental endorheic base levels

Type of base level	Numbers on Fig. 8.1	Name	Aridity	Saline lakes, salt pans, brines	Rifts, troughs intermountain basins	Young and/ or current tectonic deformation	Young and/or current seismicity	Young and/or current volcanism	Young and/or current hydrothermal features	Gulf or embayment in the nearby coastline	Existence as a paleo-base level	Existence of a paleo-sea arm	Shallow groundwater divide between base level and the sea
Marine base level	1	Mediterranean Sea									X	X	
	2	Black Sea									X	X	
Continental base level above sea level	3	Lake Chad	X								X		
	4	Great Salt Lake, USA	X	X	X	X	X		X		X		
	5	Andean Altiplano Basin	X	X	X				X		X		
	11	Aral Sea, Russia	X	X							X	X	
	21	Mono Lake	X	X	X	X	X	X	X		X		
	6	Main Ethiopian Rift Lakes			X	X	X	X	X		X		
Continental base level below sea level – distant from the sea	7	Great Artesian Basin, Lake Eyre, Australia	X	X	X	X	X		X	X	X		X
	8	Death Valley, USA	X	X	X	X			X		X		
	9	Central Asia Closed Basins	X	X	X	X			X		X		
	10	Caspian Sea	X	X							X	X	
Continental base level below sea level – close to the sea	12	Dead Sea Rift, Israel and Jordan	X	X	X	X	X	X	X	X	X	X	X
	13	Afar Depression	X	X	X	X	X	X	X	X	X	X	X
	14	Qattara Depression Egypt	X	X							X		X
	15	Salton Trough, USA	X	X	X	X	X	X	X	X	X	X	X
	16	Chotts of Tunisia and Algeria	X	X	X	X				X	X	X	X
	17	Sirte Basin, Sabkhat Ghuzayil, Libya	X	X	X	X				X	X	X	X
	18	Lago Enriquillo, Dominican Republic		X	X	X	X			X	X	X	X
	19	Argentine Salinas	X	X									
	20	Sebkha Paki Tah, Morocco	X	X									

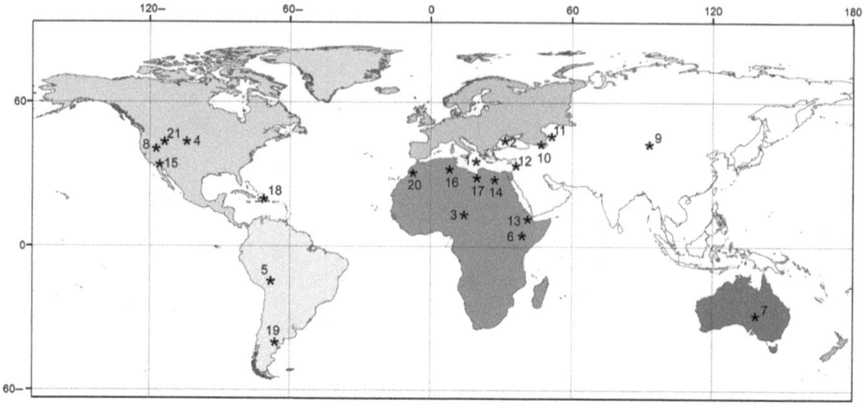

Fig. 8.1 Location map of the known continental endorheic base levels in the world. *Numbers* refer to sites as detailed in Table 8.1

The location of the relatively famous and described endorheic basins in the world is shown on Fig. 8.1 and listed below. The different continental base-levels are grouped into three categories, namely base-levels above sea level (Chap. 9) and base-levels below sea level, both distant (Chap. 10) and close to the sea (Chap. 11).

References

Almendinger JE (1990) Groundwater control of closed-basin lake levels under steady-state conditions. J Hydrol 112:293–318

Gosselin DC, Khisty MJ (2001) Simulating the influence of two shallow, flow-through lakes on a groundwater system: implications for groundwater mounds and hinge lines. Hydrogeol J 9:476–486

Pederson DT (2001) Stream piracy revisited: a groundwater sapping solution, GSA Today 11:4–10

Person M, Prasenjit R, Wright H, Gutowski Jr. W, Ito E, Winter T, Rosenberry D, Cohen D (2007) Hydrologic response of the Crow Wing Watershed, Minnesota, to mid-Holocene climate change. Geol Soc Am Bull 119:363–376

Rodriguez-Rodriguez M, Moral F, Benavente J (2008) Hydrogeological characteristics of a groundwater-dependent ecosystem (La Lantejuela. Spain). Water Environ J 22:137–147

Smith AJ, Donovan JJ, Ito E, Engstrom DR (1997) Ground-water processes controlling a prairie lake's response to middle Holocene drought. Geology 25:391–394

Urbano LD, Person M, Hanor J (2000) Groundwater–lake interactions in semi-arid environments. J Chem Explor 69–70:423–427

Urbano LD, Person M, Kelts K, Hanor JS (2004) Transient groundwater impacts on the development of paleoclimatic lake records in semi-arid environments. Geofluids 4:187–196

Chapter 9
Current Continental Base-Levels Above Sea Level

9.1 General

Among the best known current continental base-levels, above sea level, which existed already in the past as paleo base-levels, are the Lake Chad Basin, the Great Salt Lake Basin, the Andean Altiplano Basin, the Aral Sea Basin, the Mono Lake Basin and the Main Ethiopian Rift (MER) lakes basins which will be discussed below (locations 3, 4, 5, 11, 21 and 6 on Fig. 8.1 respectively).

All of these are distant from the sea. The lakes within these base levels are typical of a saline environment, except of Lake Chad and the MER lakes. There exist many other examples of similar, but less known, base-levels, i.e., the Fuenta de Piedra basin in Spain (Kohfahl et al. 2008), which are not discussed herein.

The detailed history and characteristics of these continental base-levels above sea level, are described below as follows:

9.2 Lake Chad Basin

Lake Chad is an endorheic landlocked continental lake in central Africa (Fig. 8.1). The surface area of the lake's drainage basin is some 2.5 million km^2 (Leblanc et al. 2006). The lake serves as a terminal base-level to surface flow and the main rivers that feed the lake from the south are the Chari and Logone rivers (Fig. 9.1). It is important to note, for further discussion of the lake's salinity (see below) that the lake itself is not the lowest topographic depression within the basin. The Bodele Depression, to the north east of it, is topographically considerably lower than the lake's level.

The entire basin is characterized by a semi-arid climate with average rainfall of some 300 mm/y and an average evaporation as high as 200 cm/y (Isiorho and Matisoff 1990). The lake water is relative fresh between 120 and 320 mg/l, despite the fact of its being a terminal lake, located in a semi arid environment. This

U. Kafri and Y. Yechieli, *Groundwater Base Level Changes and Adjoining Hydrological Systems*,
DOI 10.1007/978-3-642-13944-4_9, © Springer-Verlag Berlin Heidelberg 2010

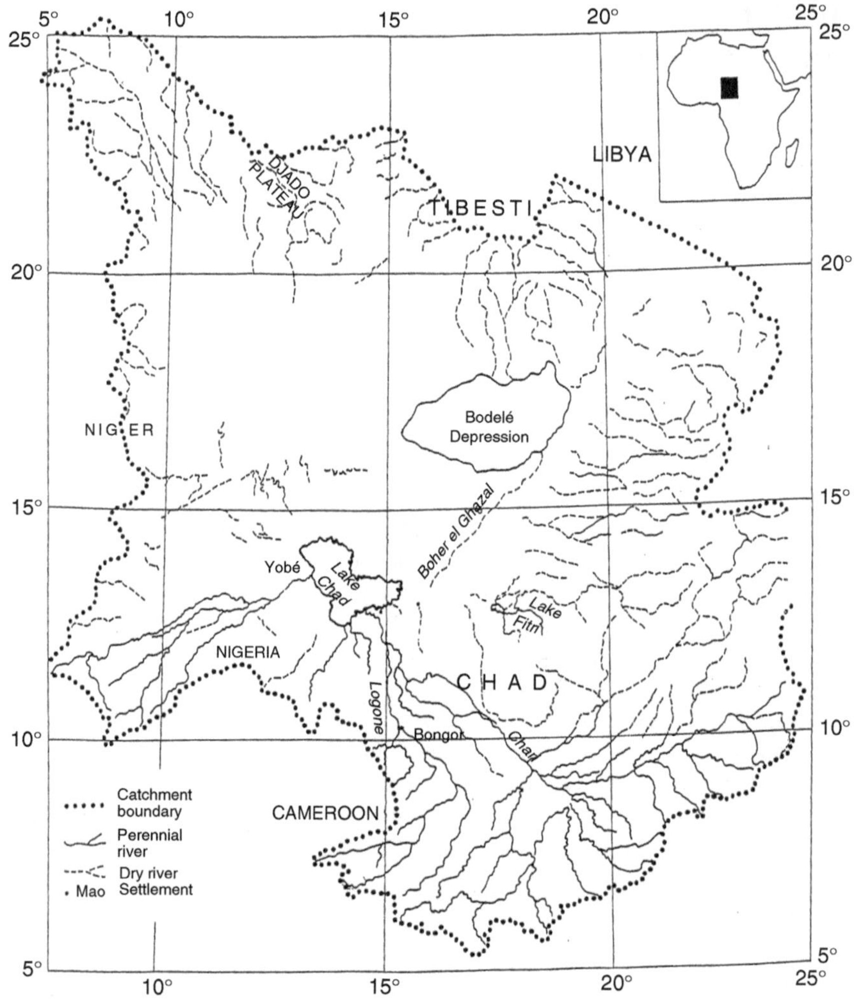

Fig. 9.1 Location map of the Chad Basin [after Kafri and Arad (1972)]

surprising situation attracted the interest of many authors (i.e., Kafri and Arad 1972; Arad and Kafri 1975) and is, thus, discussed below.

The Chad Basin was also tackled by several authors for its paleo climates and paleohydrology. Contrary to the present semi-arid climate and the resultant shrunk dimensions of the lake, wetter climates prevailed in central Africa in the past. This is evidenced by higher paleo shore lines, indicating periods of more intense surface drainage as well as a higher recharge rate to the aquifers, which enables to delineate the extent of the paleo continental landlocked lakes that occupied central Africa at those periods.

A humid climate prevailed in the region in late Quaternary times prior to the last glacial maximum (LGM) evidenced by higher recharge to the aquifers (Edmunds et al. 1999) and by an expanded Mega Lake Chad of higher lake levels (Leblanc et al. 2006). The LGM period, in this area, was an arid period, characterized by diminished surface flows and recharge to the aquifers, which resulted in the shrinkage of the previous expanded water bodies (Edmunds et al. 1999; Gasse 2000). During the wet mid Holocene, following the rise of the regional water tables, the inland Mega Lake Chad occupied again vast areas in central Africa (Gasse 2000), as much as 300,000 km^2 according to Ghienne et al. (2002), 360,000 km^2 according to Drake and Bristow (2006) or 340,000 km^2 according to Leblanc et al. (2007) (Fig. 9.2). The peak of Mega Lake Chad is estimated at around 7,500–7,000 year BP, according to Drake and Bristow (2006), or 7,700–5,500 year BP according to Leblanc et al. (2006).

Fluctuations of the paleo Lake Chad levels continued until present time, effected by climate changes, between droughts and inundations, as well as by anthropogenic influence related to water exploitation (Birkett 2000; Leblanc et al. 2007).

The groundwater regime of the basin and its groundwater chemistry were studied among others by Barber (1965), UNESCO (1969), Arad and Kafri (1975), Isiorho and Matisoff (1990), Isiorho et al. (1996), Edmunds et al. (1999, 2002),

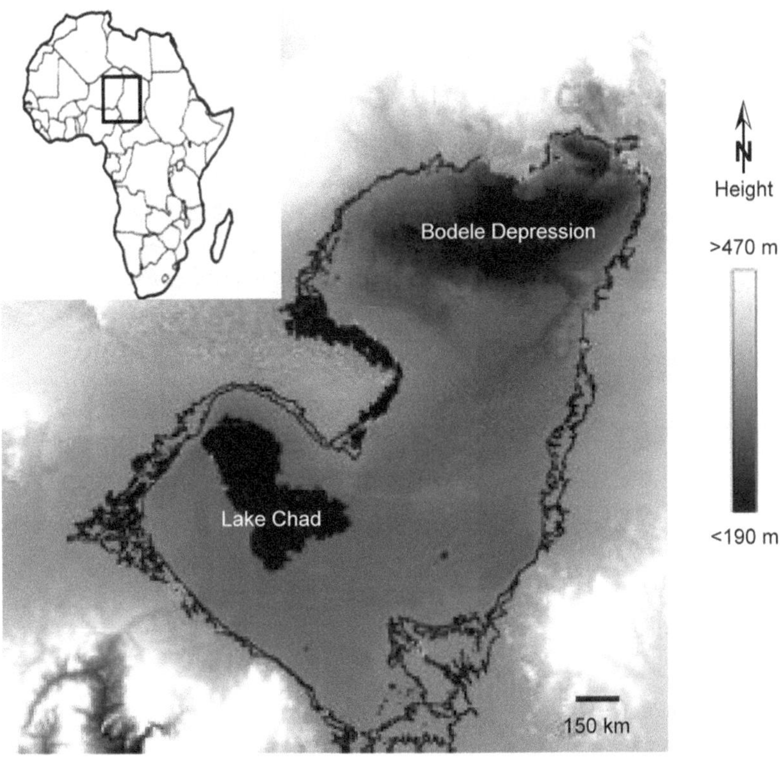

Fig. 9.2 The extension of the Megachad Lake [after Drake and Bristow (2006)]. Note that the Bodele depression is significantly lower than Lake Chad

Birkett (2000) and Leblanc et al. (2007). The hydrogeological regime of the basin, according to the above can be summarized as follows:

The groundwater systems that drain to the lake consist of the lower confined, and the overlying shallow phreatic aquifers of the Plio-Pleistocene Chad Formation. The lower confined aquifer contains mainly paleo groundwaters, that were recharged at earlier wet periods, whereas the upper phreatic one is subjected to current recharge at the rate of few to several tens of millimeter/year. The shallow phreatic groundwater table is close to the surface, mainly in the topographic depressions, a configuration which enhances evaporation. A convergent groundwater flow pattern, of both the phreatic and confined aquifers, to the lake was introduced in a report by UNESCO (1969). Different from the convergent surface water drainage regime into the lake, and the setup described by the UNESCO report, groundwater flow and seepage from the lake area proper to the southwest was already indicated by Isiorho et al. (1996) and Matisoff (1990) and by Isiorho et al. (1996). Kafri and Arad (1972) and Arad and Kafri (1975) proposed, even earlier, a groundwater flow configuration that enables groundwater flow from the lake area both to the southwest and to the northeast, toward the Bodele Depression (Fig. 9.3). Both configurations were suggested in order to explain the enigmatic freshness of Lake Chad waters (see below).

Additional enigmatic features in the basin are the described water table depressions (Arad and Kafri 1975), piezometric depressions, concentric sinks, or the ones determined as "hollow aquifers", that might attain a depth of up to 40 m, as described by Leblanc et al. (2007). The latter related this phenomenon to evaporation, despite the fact that a 40 m depth of the water table is too deep to enable evaporation and moreover, far below the root zone to enable evapotranspiration. An alternative explanation is that these sinks reflect "shortcuts" between an upper aquifer with a higher groundwater level that overlies a lower aquifer with a lower piezometric level. Such a configuration would enable groundwater flow from the upper aquifer via the lower one to the topographically lower

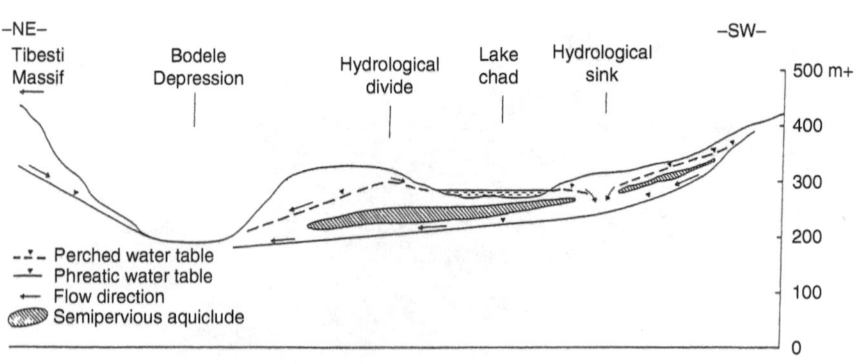

Fig. 9.3 Schematic cross section of the hydrological regime of the Chad Basin [after Kafri and Arad (1972)]

terminal base-level of the Bodele Depression (Kafri and Arad 1972; Arad and Kafri 1975) (Fig. 9.3).

The low salinity of Lake Chad, which was considered in the past as a terminal base-level situated in a semi-arid region, was regarded as a puzzle and different mechanisms were introduced to explain the phenomenon. Arad and Kafri (1975) proposed a hydrological configuration (Fig. 9.3) whereby the lake is actually not a terminal groundwater base-level, but rather a flow-through lake that recharges the groundwater system and that the actual terminal base-level is the Bodele Depression. Indeed, as mentioned before, a seepage to the aquifers was described later by Isiorho and Matisoff (1990) and Isiorho et al. (1996), claiming that this setup removes a considerable salt from the lake. The occurrence of natron and trona crusts at the bottom of the Bodele Depression is also indicative of its function as a discharge zone, supporting the above proposed model.

9.3 The Great Salt Lake Basin

The Great Salt Lake in Utah, western US, at coordinates 41°N 112°W (Figs. 8.1 and 9.4) is the fourth largest landlocked terminal lake in the world, with a surface area of about 6,000 km^2 at its historic high elevation. The basin is bounded by mountain ranges, and is part of the Basin-Range province of western US. The area was dominated since the Neogene by extension processes, and, thus, is dissected by faults that were active also during the Quaternary. Holocene neotectonic activity of fault displacement and paleoseismicity occurred since about 13.5 ka, based on high resolution seismic stratigraphy (Colman et al. 2002). The fault system is responsible for the division of the Salt Lake Basin into the north and south basins (Kowalewska and Cohen 1998). The present day lake is a saline remnant of the Pleistocene paleo-Lake Bonneville.

The palaeolimnology of the Great Salt Lake Basin was described in several studies. The basin was characterized, since the Neogene some 5 Ma ago, as a marsh land and shallow lacustrine environment. Only in the middle Pleistocene, a considerable expansion of lakes occurred with different timing and histories between the northern and the southern basins (Kowalewska and Cohen 1998). Paleo-Lake Bonneville was the predecessor of the present Great Salt Lake. The lake level history of Lake Bonneville, since 200 ka, was studied, based on $^{87}Sr/^{86}Sr$ ratios and radiocarbon ages (i.e., Hart et al. 2004). The fluctuations of paleo-Lake Bonneville are shown on Fig. 9.5. The largest lake at its highest highstand, at 1,550 m asl, existed between 15.5 and 14.5 ka. At that time span the lake covered an area of some 50,000 km^2 (Fig. 9.4). The subsequent Holocene lake levels were studied based on radiocarbon dating of nearshore sediments (Oviatt et al. 2005). Since the Holocene, the lake level dropped and its dimensions were reduced to the present Great Salt Lake.

The size of the current Great Salt Lake varies according to annual and seasonal fluctuations in evaporation and fresh water input to the basin from precipitation as

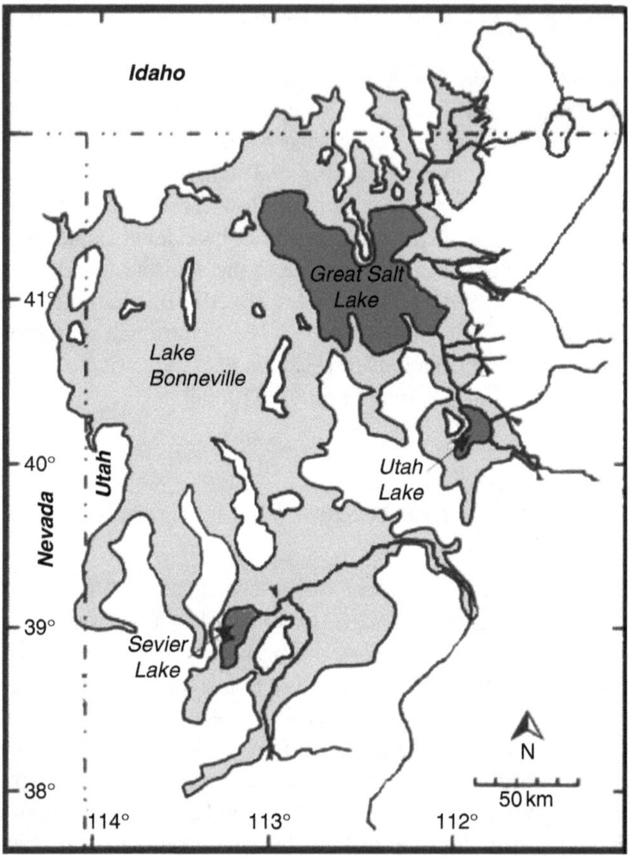

Fig. 9.4 Map of the Great Basin including Lake Bonneville and Great Salt Lake [after Hart et al. (2004)]

well as surface water and groundwater (Pedone and Folk 1996). The main fresh water input to the basin, based on noble gases studies of groundwater, is groundwater recharged in the bordering high mountains that recharges the basin through the mountain fronts (Manning and Solomon 2003).

The current lake level is around 1,280 m asl and historic fluctuations in the last century were between 1,277 and 1,284 m asl (Stephens 1990; Pedone and Folk 1996). The total salinity of the lake waters varies in time and space dependent on the balance between inflow and evaporation (Stephens 1990). The lake water is basically of the Na-chloride brine type and the total salinity is between 218 and 350 g/l in the northern basin and between 115 and 243 g/l in the southern basin (Kowalewska and Cohen 1998). The paleo-geochemistry, as well as the current geochemistry of the Great Salt Lake waters, was described by Spencer et al. (1985a, b) respectively. According to the above studies, a saline water lake existed between 30 and 19 ka followed by the expanded fresh water Bonneville paleo-lake between

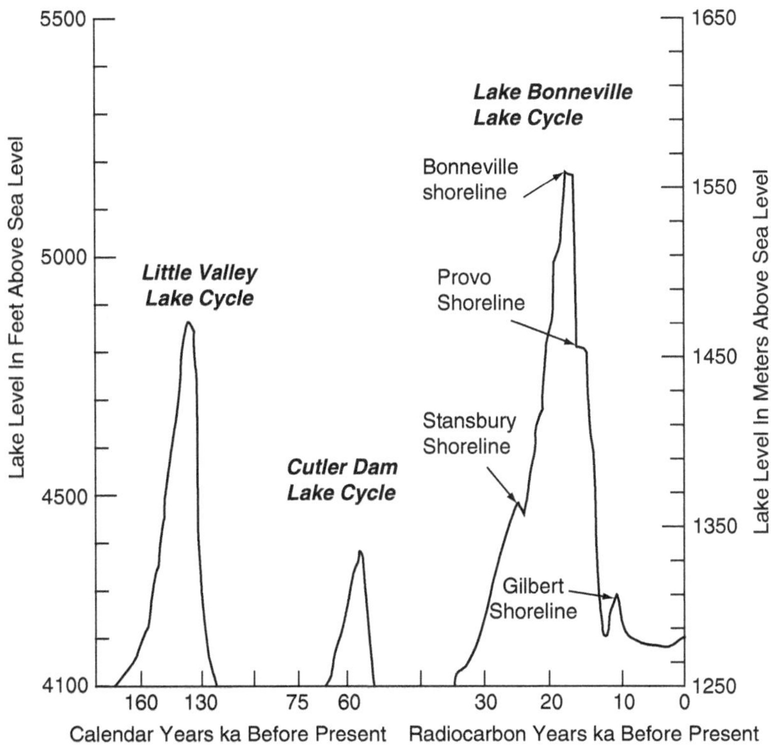

Fig. 9.5 Reconstructed Lake Bonneville levels during the last 200 ka [after Hart et al. (2004)]

19 and 15 ka. The subsequent drop of lake level resulted in a saline phase. The fresh water phase was dominated by fresh bicarbonate water input of river waters depositing carbonates. The saline phases, including the current situation, are typified by input of saline hydrothermal Na-Cl dominated springs which contributed to the salt deposition. The process was, in addition, accompanied by evaporite dissolution and diffusion of solutes into the lake bottom sediments.

The current lake is a typical playa lake situated in an arid closed basin. The combination of occupied brines of different densities, coupled with the heterogeneity in hydraulic conductivity of the system, is responsible for a complex subsurface flow pattern as modeled by Fan et al. (1997) and Holzbecher (2005). The proposed models, which are relevant to the Great Salt Lake Basin as well as to other salt lakes of closed basins, is exhibited in Fig. 6.5. The basics of the models are as follows: the basin which is bounded by hydrological divides in the mountain range drains into the center of the basin or sub-basin. Convective flows occur from the mountain front toward the basin, as well as downward density-driven brines from the center of the basin in an opposite direction. Both flows are emerging and being discharged as springs at the hinge line or margin of the basin.

9.4 The Andean Altiplano Basin

The Andean Altiplano Basin (Fig. 8.1) is a regional elongated endorheic mega basin, between the eastern Cordillera of the Andes and the western Volcanic arc, a few hundred kilometer east of the Pacific coast, and between latitudes 14°S in the north and 28°S in the south. The basin is bounded by an eastern divide which separates it from the Amazon Basin and a western one which separates it from the Pacific basins (Fig. 9.6) (Vandervoort et al. 1995). The basin appears as an elevated plateau that was uplifted to a few kilometers asl, and that was established as an internal drainage system in Miocene times, around 15 Ma (Vandervoort et al. 1995). This mega basin occupies several lakes, and salt pans (Salars) among which are the lakes of Titicaca and Poopo and the salt pans (salars) of Coipasa, Uyuni and Atacama (Fig. 9.6).

Another parallel system of a closed terminal basin also exists, more to the west, between the high Andean Cordillera and the Coastal Cordillera, namely the Pampa del Tamarugal (Fig. 9.7) (Magaritz et al. 1990; Aravena 1995).

The present climate of the discussed area, regarding rainfall and evaporation, is mostly arid to semi-arid in the central and southern parts of the mega basin and more humid in the northern Chilean part.

The paleoclimate and paleohydrology of this basin can be reconstructed based on data related to the different sub-basins, namely the present day salt pans (salars) and lakes, and the different paleo-lakes. These data include the following: lake levels and hydrological data; types of basin deposits, whether fresh water or evaporites; diatoms as salinity indicators; surface water, lakes, rain and groundwater salinity, geochemistry and isotopic composition as well as geomorphological features.

Among the several studies that analyzed the paleo-climate and hydrology are: Risacher and Fritz (1991), Grosjean (1994), Grosjean et al. (1995), Coudrain-Ribstein et al. (1995), Seltzer et al. (1998), Sylvestre et al. (1999), Valero-Garces et al. (2000, 2003), Cross et al. (2001), Rech et al. (2002), Sylvestre (2002), Fritz et al. (2004), Godfrey (2006) and Ekdahl et al. (2008). A summary of the paleoclimatic and paleohydrologic setups that prevailed in the Altiplano mega-basin, based on the above is as follows: Since its formation as an uplifted closed endorheic basin in the Neogene, the basin was subjected to climatic changes from humid periods to dry ones. These changes affected the water budget (precipitation or inflow minus evaporation) of the basin and its sub-basins, and are thus reflected in their hydrological setups. Humid periods are indicated by high lakes and groundwater levels, and by fresh water sediments in the basin fill (clays, carbonates, detrital material), whereas dry periods are indicated by low lakes and groundwater levels, by saline waters and by accumulation of thick sequences of evaporites, such as gypsum and halite.

The large Salar de Atacama exhibits salt deposition, reflecting dry climates since 5–10 Ma (Godfrey 2006). Salt was also being deposited between 325 and 53 ka and only around 50 ka a change is observed to a more humid climate, with more

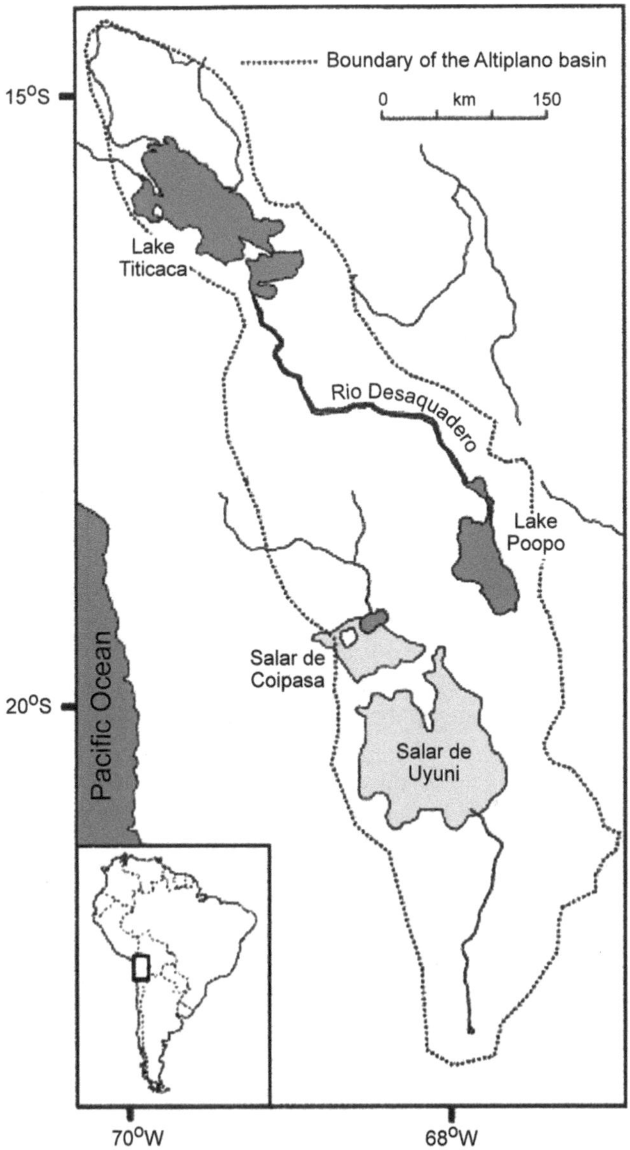

Fig. 9.6 Map of the Altiplano showing the different lakes, salars and major rivers

abundant clay sequences, as compared to the evaporite ones (Fritz et al. 2004). It should be noted that the Atacama desert is an extremely arid system, with almost no precipitation.

The paleo-lake Tauca, which preceded the present day Poopo Lake and the Coipasa and Uyuni salars, exhibits high lake levels (reflected by mud units) between 46–36 and 26–15 ka (Fritz et al. 2004). Paleo-Lake Coipasa also exhibits

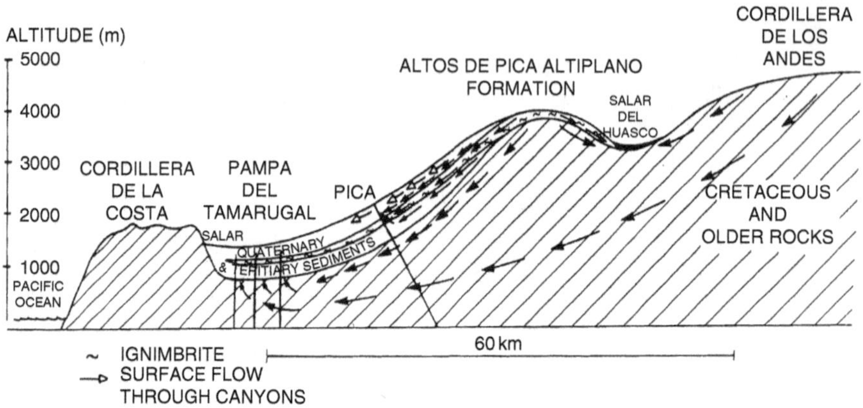

Fig. 9.7 A hydrogeological cross section from the Andes through the endorheic basins of the Salar del Huasco and the Pampa del Tamarugal, northern Chile [after Magaritz et al. (1990)]

a high lake level between 21 and 17 ka (Sylvestre 2002). Paleo-Lake Tauca maintained a high lake level in the late Glacial to early Holocene period, between roughly 16 and 9 ka, with some short time spans of lower lake levels, prior to its desiccation (Risacher and Fritz 1991; Grosjean et al. 1995; Cross et al. 2001; Rech et al. 2002). The Holocene exhibits a variability of climates (Ekdahl et al. 2008), and frequent climatic changes in the last millennia and the last 500 years (Valero-Garces et al. 2000, 2003).

The connection between the different lakes and paleo-lakes and salars in the northern and central parts of the mega-basin (Fig. 9.6) was described among others by Guyot et al. (1990), Sylvestre et al. (1999) and Fritz et al. (2004). Lake Titicaca, in the north, is a fresh water body (salinity around 1 g/l) that drains to the south through the river Rio Desaguadero to Lake Poopo. The southern outlet of Lake Titicaca constitutes a very shallow sill, a few meter below the lake level (Cross et al. 2001). During a shallow lake level phase, below that sill, as happened in early-mid Holocene (Seltzer et al. 1998), the closure of the lake as a result ended up as its functioning as an endorheic lake. During high lake levels, the lake had overflowed and drained to Lake Poopo and to the combined terminal Coipasa and Uyuni salar, which form the biggest salt pan in the Andes.

Due to the present day general aridity of the area proper, it is not subjected to direct recharge from precipitation, but rather from groundwater flow that is being recharged on the surrounding high Cordilleras of the Andes. The altitude dependent depleted stable isotopic compositions of the different groundwater types are indicative of the elevations of their related recharge zones as shown by Magaritz et al. (1990) and Aravena (1995). Groundwater flows, recharged in the mountains, converge and are discharged into the terminal lakes and salars (Magaritz et al. 1990; Aravena 1995), as happens in other similar regions in the world. A cross section from the high Cordillera through the endorheic basin of Pampa del Tamarugal in northern Chile (Fig. 9.7) (Magaritz et al. 1990) displays the hydrological configuration that includes the recharge zones, the groundwater flow directions and the terminal base-levels and discharge zones with the divides in between. In this same area, paleo climatic and tectonic controlled groundwater level declines were inferred from deep canyons incision, created by groundwater sapping, as described by Hoke et al. (2004).

The evolution of groundwater chemistry is noticed from the periphery toward the base-levels and discharge zones. Fresh groundwater picks up salinity through water rock interaction, evaporation and dissolution of evaporites, deposited in the paleo lakes and salars as evidenced, for example, from bromine geochemistry of brine and of deep salt crusts in the Uyuni Salar (Risacher and Fritz 2000, 2009). An additional contribution to salinity are emerging hydrothermal Ca-Cl brines from depth, associated with active faulting and volcanism, as the ones described from the southwestern margins of the Salar de Atacama, a phenomenon described also from other closed basins in the world (i.e., Death Valley) (Lowenstein and Risacher 2009) (see also Sect. 6.3.1).

It is suggested herein that in the case of the endorheic Altiplano mega basin, as elsewhere in the world, the lakes and salars are also groundwater dependent and serve as groundwater base-levels and discharge zones. Thus, any changes in base-level elevation, whether climatically or tectonically controlled, should have also affected the groundwater levels rises or declines.

9.5 The Aral Sea Basin

The Aral Sea endorheic terminal basin (Fig. 8.1), located at coordinates 45°N 60°E (Fig. 9.8) was in 1960 the fourth largest inland water body in the world, prior to its shrinkage to a tenth of its original size by 2007 (Micklin and Aladin 2008).

The Aral depression existed already in the Late Neogene. Since then, a number of transgressions of the Caspian Sea reached as far as the Aral Sea basin. A direct marine connection between the Caspian Sea and the Aral Sea Basin existed about 3 Ma (Boomer et al. 2000). Three major regressions of the Aral Sea occurred around 2 Ma, in the LGM around 20 ka, both associated with the Caspian Sea lowstands, and in the Late Holocene. All three regressions resulted in salt deposition due to evaporation of the water body (Boomer et al. 2000).

The most recent water input to the depression took place by the Syr Darya River and later, in the beginning of the Holocene, by the capture of the Amur Darya River. Since then, the sea is fed by these two rivers (Micklin 1988). In the last ten millennia, the Aral Sea level fluctuated by over 40 m as a result of climate changes and thus the inflows of the feeding rivers, and in addition, due to anthropogenic water withdrawal for irrigation in the last century (Micklin 1988).

In the 1960s, the average sea level was 53.4 m asl and its surface area was some 68,000 km^2. Enhanced water withdrawal for irrigation, which increased the evapotranspiration in the Aral Sea drainage basin, and resulted in reduced inflows from the rivers, caused a sea level drop by 23 m at a rate of 0.6 m/y (Fig. 9.9). Due to the higher evaporation/inflow ratio, the surface area of the sea was reduced to the figure

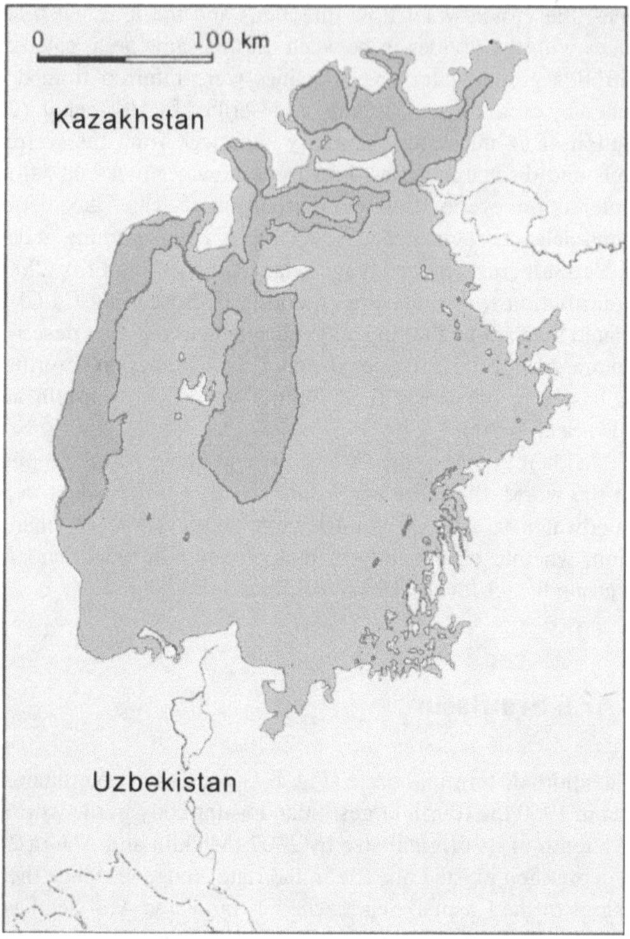

Fig. 9.8 Map of the Aral Sea. The extension of the sea in 1960 and 2008 is given by the lighter and darker colors, respectively

of 17,000 km^2 in 1985 (Fig. 9.8) (Micklin 1988, 2007; Small et al. 2001; Benduhn and Renard 2004; Peneva et al. 2004; Shibuo et al. 2007). As a result, the sea was divided into two separate water bodies, the shallow Small Aral Sea in the north and the Large Aral Sea in the south (Johansson et al. 2009). The desiccation and shrinkage of the sea was accompanied by a drastic increase of salinity from around 10 g/l in the 1960s to over 100 g/l currently. Sodium sulfate and chloride salts were deposited on the bottom of the sea, and when subsequently were exposed following the shrinkage of the sea, they were wind blown to long distances, causing environmental damage (Micklin 1988, 2007). The salination and pollution of the groundwater systems were caused mainly by agricultural irrigation Johansson et al. (2009).

The drop of the sea level has been accompanied by the lowering of the piezometric heads and by reduction of groundwater flow in the adjoining groundwater

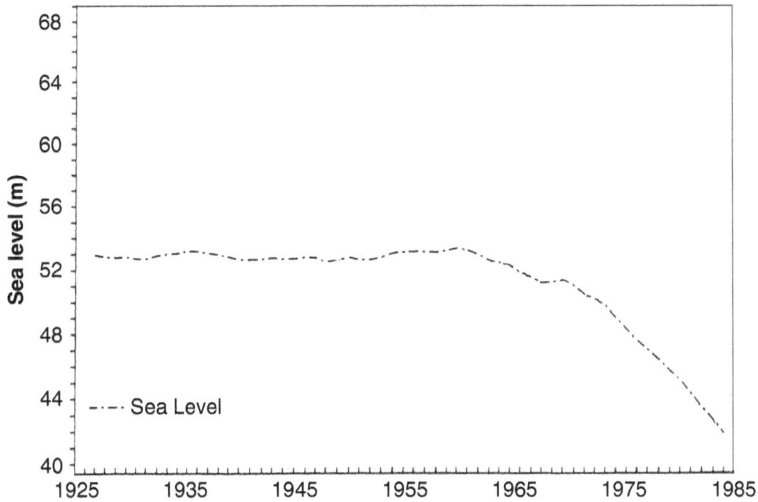

Fig. 9.9 The decline of the Aral Sea level between 1925 and 1985 [modified after Micklin (1988)]

systems. The decline of the adjoining groundwater levels was found as far as tens of kilometers away from the sea (Micklin 1988). Calculations, based on water balances, salt balances and hydraulic gradients carried out by Jarsjo and Destouni (2004), have yielded the following results regarding groundwater flow to the sea as a result of its level drop: As a result of the hydraulic gradient changes, the groundwater flows increased in the northern Small Aral Sea, and remained the same for the Large Aral Sea in the south. Thus, the groundwater component, as a part of the inflow to the sea, increased dramatically in comparison to the surface water inflow component.

9.6 The Mono Lake Basin, California

The Mono Lake, California, USA, at coordinates 38°N 119°W, (Fig. 9.10) is a perennial endorheic closed basin, and a saline lake at an elevation of about 1,950 m asl. The lake, with a surface area of some 180 km^2 and a maximum depth of 48 m, is located in a semi-arid region in the rain shadow of the Sierra Nevada ranges of western USA. It is characterized by an average annual direct precipitation, on the lake area proper, below 300 mm and an annual evaporation around 1.2 m. Precipitation over the surrounding mountain ranges is considerably higher.

The Mono Lake is located within the paleo-Mono Basin, which is the highest of the closed basins in the Basin and Range Province of western USA (Reheis et al. 2002). The Mono Basin is 3–4 Ma old and it contains up to 2,100 m thick sequence of late Pliocene and younger sediments and volcanic deposits (Gilbert et al. 1968). There is no evidence of evaporite deposits in the basin over at least the last 700 ky

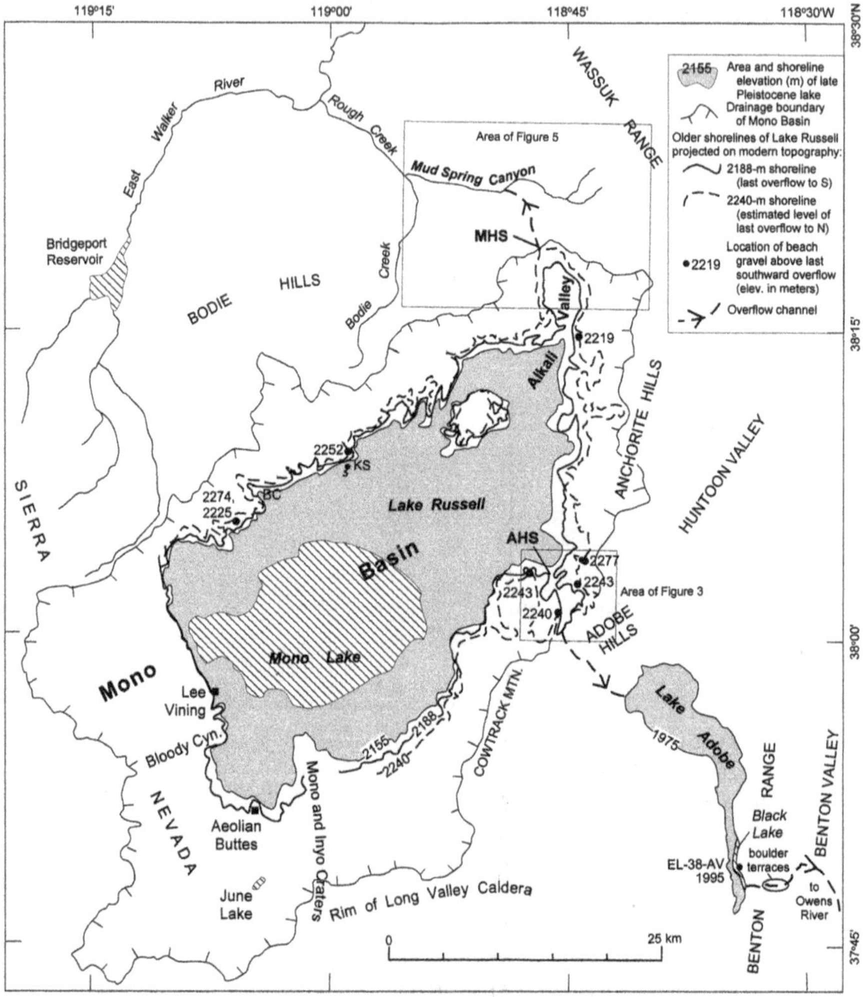

Fig. 9.10 Map showing the Mono Basin [after Reheis et al. (2002)]

(Gilbert et al. 1968). The Pleistocene paleo-Lake Russell, that occupied part of the basin (Fig. 9.10) attained altitudes between 2,205 and 2,280 m asl, far above the present Mono Lake level. During the late glacial time the lake attained elevations of 240 m above its present day elevation (Stine 1991). During its history the Russell Lake spilled to the northeast, to the Lahontan drainage basin through the Mount Hicks spillway, and later on it spilled to the south to the Owens-Death Valley system through the Adobe Hills spillway (Fig. 9.11) (Reheis et al. 2002). It is, therefore, assumed that at least during these periods the basin served as a flow-through base-level. The shift of the above described spill points of the Russell Lake was attributed to late Tertiary and Quaternary faulting and volcanism (Gilbert et al. 1968; Bursik and Sieh 1989; Reheis et al. 2002). Since the late Quaternary, the lake

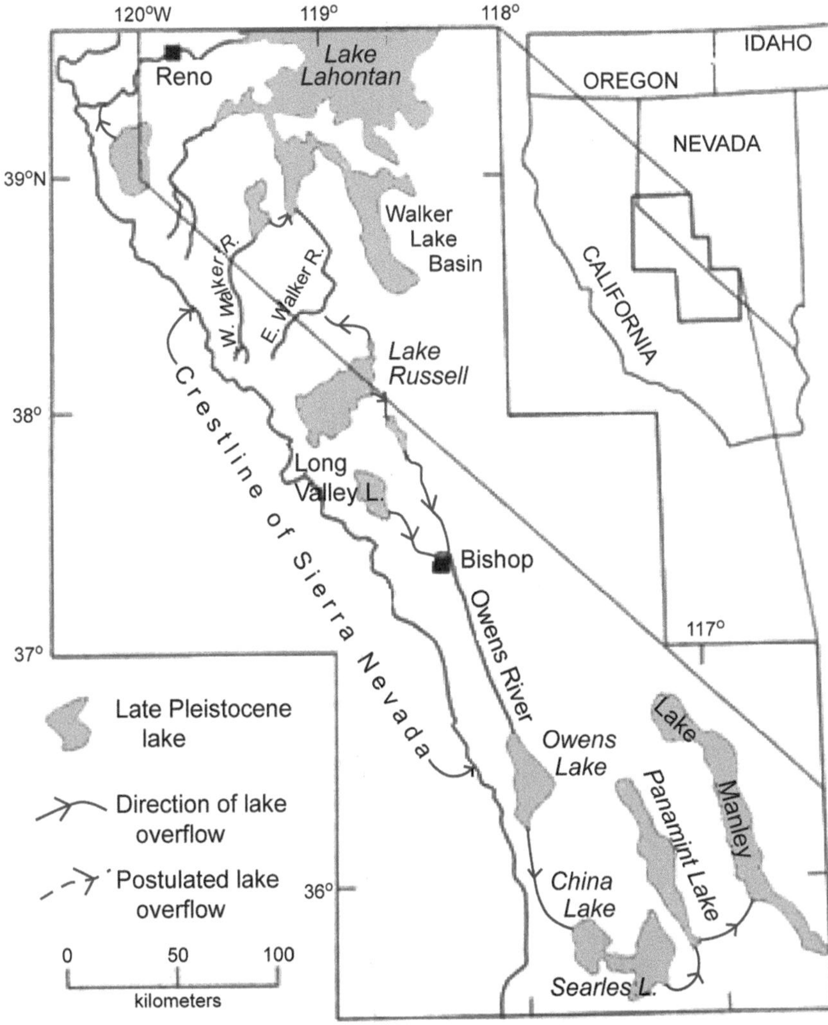

Fig. 9.11 Map showing the extent of lakes during the late Pleistocene, and the overflow connections along the eastern Sierra Nevada [after Reheis et al. (2002)]

level dropped to the present level of its remnant Mono Lake level. Late Holocene and historic lake level fluctuations, between 1,945 and 1,980 m asl, due to climatic changes, based on shoreline elevations, were reported by Stine (1990).

As part of the Basin and Range province, the Mono Basin was subjected to neotectonic movements and to current considerable slip rates along faults, as was also monitored by satellite-based geodetic data (i.e., Bursik and Sieh 1989; Dixon et al. 1995; Reheis and Dixon 1996). Young and current seismicity (i.e., VanWormer and Ryall 1980) and volcanism (i.e., Sieh and Bursik 1986) were also reported from the discussed region.

The present Mono Lake is fed by three perennial drainages from the Sierra Nevada Mountains in the southwest that contribute most of the surface water to the lake. Part of the surface flow recharges the groundwater system at the margins of the lake. Convergent groundwater flow to the lake, that serves also as a groundwater base-level, is evident from the water table map (Fig. 9.12) (Lee 1969; Rogers and Dreiss 1995a). It is noticed that the groundwater flow gradient is rather steep in the rainy intake areas of the Sierra Nevada in the southwest and is mild in the more arid sector of the northeast, as expected.

The total dissolved solids concentration of the lake's water is about 90,000 ppm whereas the basin sediments and the aquifer underneath the lake contain groundwater with a salinity exceeding 18,000 ppm in several areas (Rogers and Dreiss 1995a). The groundwater salinity of the Mono Lake area (Fig. 9.13) is in agreement with the groundwater flow pattern as shown on the water table map (Fig. 9.12). Namely, a fresh water input from the intake areas in the southwest and west and a rather saline water input from the more arid areas in the east and northeast and from underneath the center of the lake. Water issuing as thermal springs in the Paoha Island, that is situated in the center of the lake, exhibit concentrations as high as 26,000 ppm. These are interpreted as a mixture between lake water and waters from a deep source, which is also supported by mantle source helium isotopes content of this water type (Rogers and Dreiss 1995a; Clark and Hudson 2001). Ascending groundwater from underneath the lake is also responsible for the formation and deposition of the famous calcium carbonate tufa deposits and towers that crop out in the lake's bottom and above the lake's water level.

The high groundwater salinities, of up to 100,000 ppm, found along the shallow exposed northeast shorelines (Fig. 9.13), are attributed to high evaporation of the

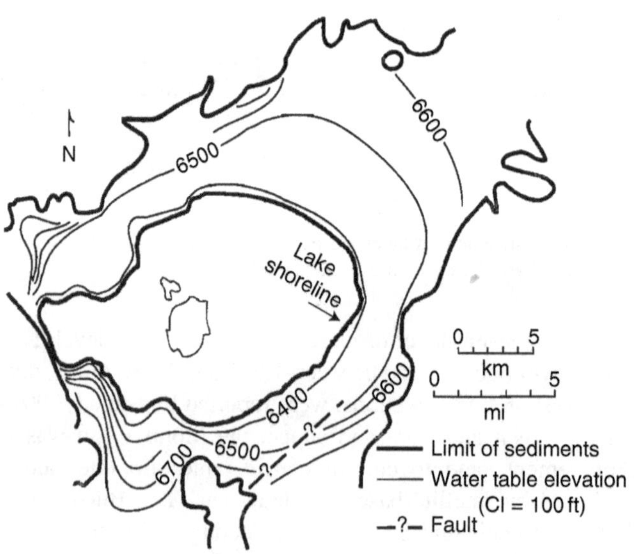

Fig. 9.12 Mono basin water table elevation map [after Lee (1969)]

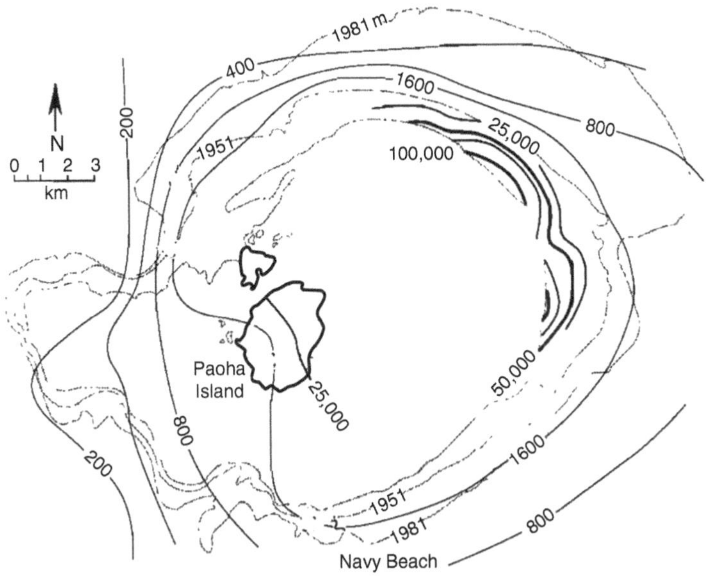

Fig. 9.13 Mono Lake groundwater salinity (in ppm) [after Lee (1969)]

playa surface and to upwelling of deep saline groundwater there (Rogers and Dreiss 1995a, b). The location of the fresh/saline groundwater interface around the lake margins, as described by Rogers and Dreiss (1995a), is controlled by the following: Higher recharge rates from the Sierra Nevada in the southwest push the interface in this sector of the lake far beneath the lake, whereas the interface lies around and below the shorelines in most of the rest of the lake. On the low recharge northeast shorelines, the interface may lie outside the lake edge and the discharge zone being situated below the playa.

The inter-relationship between the Mono Lake levels and the salinity of both the lake water and the groundwater underneath was modeled and discussed in detail by Rogers and Dreiss (1995b). According to them, at low lake levels the high water salinity causes solute loss via diffusion into the bottom lake sediments and via advection, mostly through faults and fractures, into the groundwater system. At higher lake levels, the shorelines discharge zones move toward the basin edge and the saline groundwater mass is subsiding, drawing solutes from the lake into the basin sediments.

Falling lake levels again constricts the saline groundwater beneath the lake, forcing saline waters into the lake and thereby increasing its salinity. Long-term paleo and shorter-term historical salinity and lake levels or lake volume time series (Figs. 9.14 and 9.15) indeed show the inverse relationship between lake levels and lake salinities as described before. High lake levels were found to be associated with lower lake water salinity (Fig. 9.16) and vice versa (Stine 1987; Stine 1991; Rogers and Dreiss 1995b).

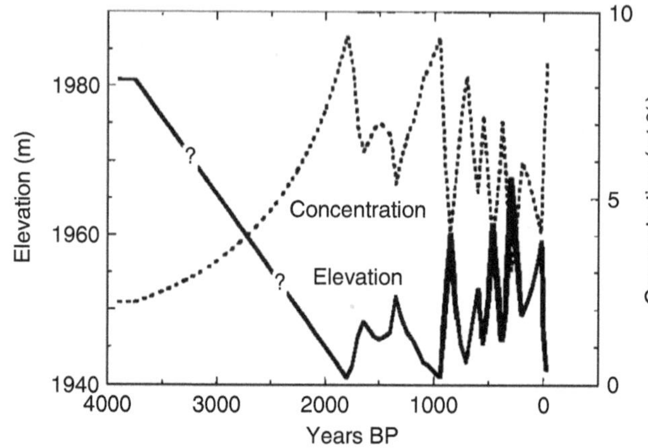

Fig. 9.14 Paleo Mono Lake levels and salinities [after Stine (1987)]

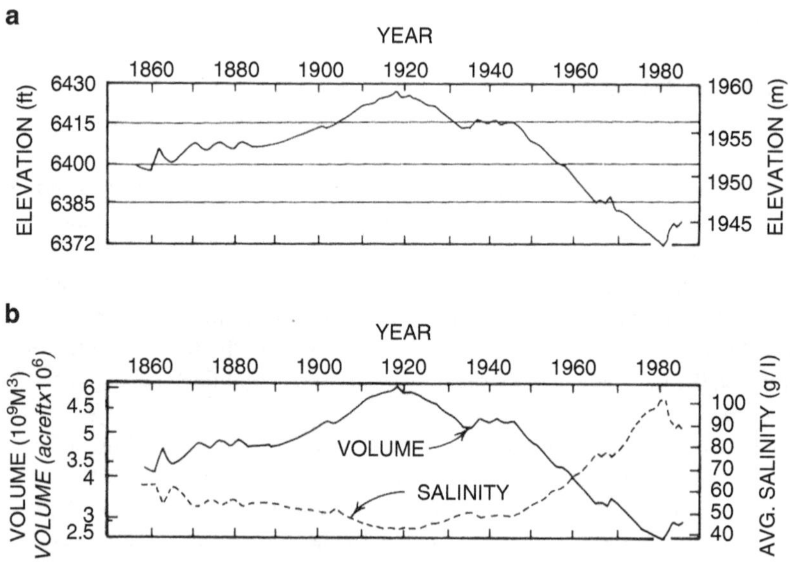

Fig. 9.15 Historical Mono Lake (**a**) levels, (**b**) volumes and salinities [after Stine (1991)]

9.7 The Main Ethiopian Rift Basin

The Main Ethiopian Rift (MER) Basin is an elongated, SSW–NNE trending feature, considerably above sea level, which contains a chain of closed, or partly connected basins and lakes that are aligned along it, roughly between coordinates 7°N 38°E and 11°N 41°E (Fig. 9.17). Arid to semi-arid conditions prevail in part of the discussed region whereas on the bounding highlands the climate is more humid.

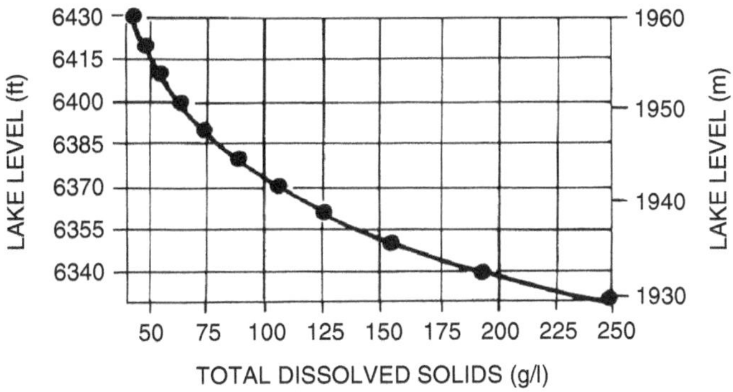

Fig. 9.16 The relationship between Mono Lake levels and salinities [after Stine (1991)]

The MER is part of the East African Rift system that joins the Red Sea and the Gulf of Aden rifts to form the triple junction of the Afar Depression (Fig. 9.17). The structure, tectonics and volcanism of the MER were discussed and analyzed in numerous studies (i.e., Wolfenden et al. 2004; Bonini et al. 2005; Ayele et al. 2006; Asfaw et al. 2006; Keir et al. 2006). A concise summary of the structural framework and its development, based on the above, is as follows:

The Afar triangle, which includes the northern segment of the MER, is a plate boundary between Africa and Arabia, which exhibits the separation of both by sea floor spreading accompanied by rifting, recent tectonism, current seimicity, volcanism and hydrothermal activity. The entire system was formed after the Oligocene flood basalt magmatism at around 31 Ma. The East African Rift system propagated northward as does its MER segment. The southern MER was formed in the Early Miocene at around 20–21 Ma, whereas the northern MER segment and the Afar triple junction were formed only after 11 Ma. The opening of the central MER segment occurred only in the Pliocene, between 3 and 5 Ma.

The combined central and northern parts of the MER form an elongated depression, 80 km wide and some 700 km long, bounded by stepped escarpments, leading to the Ethiopian highlands and the Somali plateau. The depression occupies the endorheic lacustrine systems of the Ziway–Shala lake basin, including the present Ziway, Shala, Langano and Abijata lakes, at elevations between 1,578 and 1,636 m asl. The Ziway and Langano lakes drain via rivers to the Abijata Lake, whereas the Shala Lake forms a separate closed basin, fed by groundwater seepage from the other lakes (Fig. 9.18) (i.e., Ayenew 2003; Legesse and Ayenew 2006; Sagri et al. 2008).

The paleohydrology of the discussed region was described, among others, by Sagri et al. (2008) as follows: In the Pleistocene, a huge mega-lake flooded large parts of the MER, which used to overflow in certain periods to the northern Awash River basin (Legesse and Ayenew 2006). Following the arid conditions that prevailed in the Late Glacial Maximum (LGM), in the latest Pleistocene, that mega-lake started to shrink. This process was also enhanced by tectonic deformation and

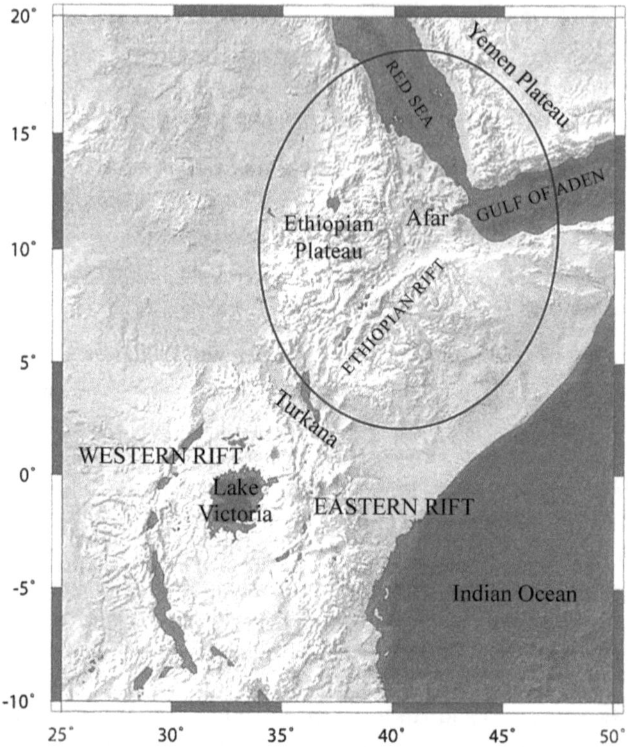

Fig. 9.17 The East African Rift system [after Yirgu et al. (2006)]. Permission to use this figure was granted by the Geological Society of London

volcanism which modified the drainage network, resulting in the capture of parts of the initial basin to the west and east. In the beginning of the Holocene, more humid conditions induced a water level rise to form the combined Ziway–Shala macro-lake basin which was subsequently separated to the present day four lakes configuration, in a more arid period.

The groundwater flow regime and pattern of the entire region, including the highland intake areas and the rift endorheic base level was described, among others, by Ayenew (2003), Ayenew et al. (2007, 2008), and Kebede et al. (2008). The aquifers basically consist of volcanic sequences or Quaternary detrital deposits. The general flow pattern is from the more rainy highlands recharge areas into the MER, partly utilizing the bounding step faults, which act as conduits. In other cases, faults are blocking and diverting the flow direction. Some of the groundwater flow issues as springs along the marginal rift step faults or grabens (Fig. 9.19). Part of the recharge to the MER aquifers takes place through alluvial channel losses to the aquifers.

In general, the groundwater flow is convergent from the highlands to the four lakes endorheic base level, as shown on Fig. 9.18. Within the base level itself, all four lakes are also groundwater dependent, fed mainly by large faults draining

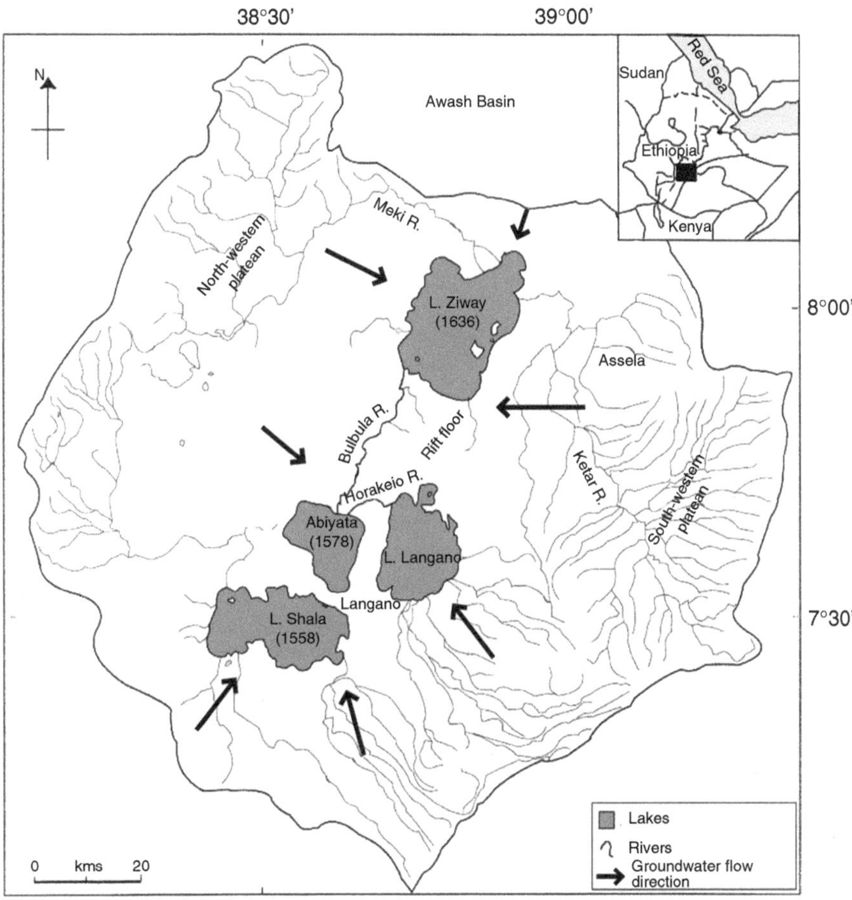

Fig. 9.18 Convergent groundwater flow into the MAIN Ethiopian Rift (MER) lakes [modified after Legesse and Ayenew (2006)]

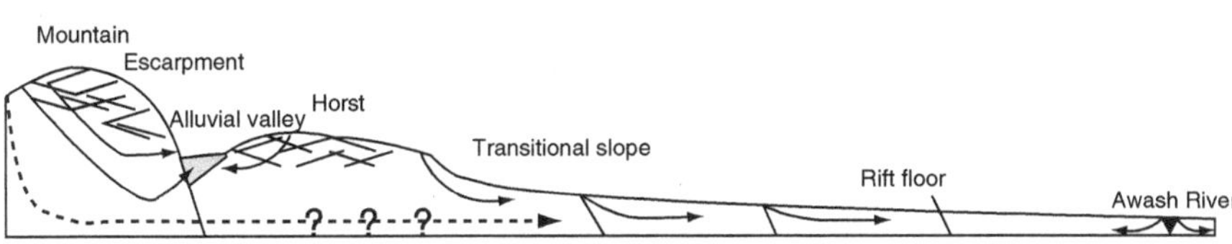

Fig. 9.19 Schematic diagram showing the groundwater flow pattern from the Ethiopian highlands into the MER aquifers [after Kebede et al. (2008)]

groundwater from the highlands, and have a subsurface hydraulic link (Ayenew 2003). The amount of subsurface groundwater outflow from Lake Awassa, for example, based on water balances, exceeds 30% of its total outflow (Ayenew et al. 2007). Thus, a seepage mechanism from one lake to another is considerable.

A somewhat similar situation prevails also in two topographically crater lakes, namely the Hayq and Ardibo lakes in the northeastern highlands of Ethiopia, outside the MER, as described by Demlie et al. (2007). Here also, the two separated crater lakes exhibit both inflow and outflow of groundwater and an hydraulic leakage link through large faults.

The salinity, geochemical affinity as well as the stable isotope composition of the lakes' waters and the different groundwater sources in the entire region were studied, among others, by Kebede et al. (1994), Ayenew (2003), Ayenew et al. (2008), and Kebede et al. (2008, 2009). A short summary of their analysis reveals the following:

A geochemical zonation of the groundwater exists, whereby the groundwater in the highlands is relatively fresh, of the Ca–Mg-bicarbonate type, with an elevation controlled, depleted stable isotope composition, as expected in these high elevations. The down-gradient, MER groundwater is already more saline, usually of the Na-bicarbonate type and locally enriched in sulfate and chloride. The waters of the lakes already exhibit an enriched stable isotope signature, different from the highlands groundwater, and typical of an evaporation-dominated system.

The salinity of the above discussed MER lakes differs from one lake to another. The salinities (total in milliequivalent/liter) of the Ziway, Langano, Shala and Abijata lakes are around 9, 34, 550 and 850 respectively (Kebede et al. 1994). The highest salinity of Lake Abijata has to do with its being the terminal one among the lakes and also due to its recent shrinkage related to climate changes, but mostly following man-made water abstraction and pollution upstream. The drop of the lake level was accompanied by an increase of its salinity (Ayenew 2002; Legesse and Ayenew 2006).

The impact of human activity on the above discussed lake levels is of great concern and, thus, was discussed in various studies (i.e., Ayenew 2002; Alemayehu et al. 2006; Legesse and Ayenew 2006). The level of Lake Abijata dropped drastically, up to 6 m, in the eighties (Fig. 9.20), and its size was reduced recently

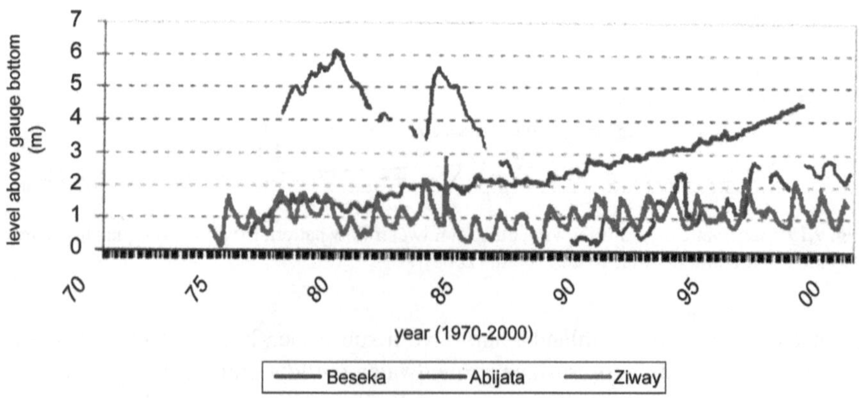

Fig. 9.20 Recent lake level changes within the MER [modified after Alemayehu et al. (2006)]

mostly due to intensive water abstraction from the lake for irrigation and soda ash production. At the same time, the levels of the Ziway and Langano lakes, with a limited human influence, exhibit no considerable changes.

The endorheic Lake Beseka, at coordinates 8°N 39°E, which is to the north of the Ziway–Shala basin, on the other hand, considerably expanded at the same time and its level has risen by some 4 m. This was attributed to both percolation from irrigated areas upstream the lake and to the rise of the Awash River level following a construction of a dam on the river. Both processes contributed to the groundwater level rise and increased groundwater contribution to the lake, whereby raising its level (Alemayehu et al. 2006). An additional input to the lake was suggested to be the increased discharge of hot springs that is interpreted as responsible to the rise of the lake level (Goerner et al. 2009). The recent behavior of the discussed lakes' levels is shown on Fig. 9.20.

To conclude, the MER lacustrine system is regarded as an endorheic basin, regarding the regional surface drainage network and the convergent groundwater flow pattern. Still, despite the fact that the lakes, within and outside the MER, are topographically closed basins, some of them are also groundwater dependent and interconnected by subsurface flow. The faults that act as conduits play an important role in forming this interconnection. Thus, the separate closed lakes practically fall in to the category of flow-through lakes, rather than the terminal endorheic ones. This is also supported by the observed considerable downward leakage from the lakes to the groundwater system, as well as to the relatively freshness of their water despite their evaporative regime.

When dealing with the entire Ziway–Shala basin as one entity, it is regarded as a combined terminal endorheic base level, based on the surface and groundwater flow patterns. Here again, one can speculate that due to the relative low water salinity, it is practically not terminal, whereby a downward leakage exists to deeper aquifers and flow via them, also utilizing the fault system, along the MER, to the ultimate Afar endorheic base level in the north (see also Sect. 11.3). In such a case, the entire system can be described as a flow-through system leading down to the real ultimate base level.

References

Alemayehu T, Ayenew T, Kebede S (2006) Hydrogeochemical and lake level changes in the Ethiopian Rift. J Hydrol 316:290–300

Arad A, Kafri U (1975) Geochemistry of groundwaters in the Chad Basin. J Hydrol 25:105–127

Aravena R (1995) Isotope hydrology and geochemistry of northern Chile groundwaters. Bull Inst Fr Etudes Andines 24:495–503

Asfaw LM, Beyene H, Mkonnen A, Oli T (2006) Vertical deformation in the Main Ethiopian Rift: leveling results in its northern part, 1995–2004. In: Yirgu G, Ebinger CG, Maguire PKH (eds) The Afar volcanic province within the East African Rift System. Geological Society of London Special Publications, London 259:185–190

Ayele A, Nyblade AA, Langstone CA, Cara M, Leveque JJ (2006) New evidence for Afro-Arabian plate separation in southern Afar. In: Yirgu G, Ebinger CG, Maguire PKH (eds) The Afar

volcanic province within the East African Rift System. Geological Society of London Special Publications, London 259:133–141
Ayenew T (2002) Recent changes in the level of Lake Abiyata, central main Ethiopian Rift. Hydrol Sci 47:493–503
Ayenew T (2003) Environmental isotope-based integrated hydrogeological study of some Ethiopian rift lakes. J Radioanal Nucl Chem 257:11–16
Ayenew T, Becht R, van Lieshout A, Gebreegziabher Y, Legesse D, Onyando J (2007) Hydrodynamics of topographically closed lakes in the Ethio-Kenyan Rift: the case of lakes Awassa and Naivasha. J Spat Hydrol 7:81–100
Ayenew T, Demlie M, Wohnlich S (2008) Hydrogeological framework and occurrence of groundwater in the Ethiopian aquifers. J Afr Earth Sci 52:97–113
Barber W (1965) Pressure water in the Chad formation of Bornu and Dikwa Emirates, Northeastern Nigeria. Bull Geol Surv Niger 35:138
Benduhn F, Renard P (2004) A dynamic model of the Aral Sea water and salt balance. J Mar Syst 47:35–50
Birkett CM (2000) Synergistic remote sensing of Lake Chad: variability of basin inundation. Rem Sens Environ 72:218–236
Bonini M, Corti G, Innocenti F, Manetti P, Mazzarini F, Abebe T, Pecskay Z (2005) Evolution of the Main Ethiopian Rift in the frame of Afar and Kenya rifts propagation. Tectonics 24:TC1007 DOI:10.1029/2004TC001680
Boomer I, Aladin A, Plotnikov I, Whatley R (2000) The palaeolimnology of the Aral Sea: a review. Quat Sci Rev 19:1259–1278
Bursik MI, Sieh K (1989) Range front faulting and volcanism in the Mono Basin, eastern California. J Geophys Res 94B:15587–15609
Clark JF, Hudson GB (2001) Quantifying the flux of hydrothermal fluids into Mono Lake by use of helium isotopes. Limnol Oceanogr 46:189–196
Colman SM, Kelts KR, Dinter DA (2002) Depositional history and neotectonics in Great Salt Lake, Utah, from high-resolution seismic stratigraphy. Sediment Geol 148:61–78
Coudrain-Ribstein A, Olive P, Quintanilla J, Sondag F, Cahuaya D (1995) Salinity and isotope dynamics of the groundwater resources on the Bolivian Altiplano. IASH Publ 232:268–276
Cross SL, Baker PA, Seltzer GO, Fritz SC, Dunbar RB (2001) Late Quaternary climate and hydrology of tropical South America inferred from an isotopic and chemical model of Lake Titicaca, Bolivia and Peru. Quat Res 56:1–9
Demlie M, Ayenew T, Wohnlich S (2007) Comprehensive hydrological and hydrogeological study of the topographically closed lakes in highland Ethiopia: the case of Hayq and Ardibo. J Hydrol 339:145–158
Dixon TH, Robaudo S, Lee J, Reheis MC (1995) Constrains on present-day basin and range deformation from space geodesy. Tectonics 14:755–772
Drake N, Bristow C (2006) Shorelines in the Sahara:geomorphological evidence for an enhanced monsoon from palaeolake Megachad. Holocene 16:901–911
Edmunds WM, Fellman E, Goni IB (1999) Lakes, groundwater and palaeohydrology in the Sahel of NE Nigeria: evidence from hydrogeochemistry. J Geol Soc Lond 156:345–355
Edmunds WM, Fellman E, Goni IB, Prudhomme C (2002) Spatial and temporal distribution of groundwater recharge in northern Nigeria. Hydrogeol J 10:205–215
Ekdahl EJ, Fritz SC, Baker PA, Rigsby CA, Coley K (2008) Holocene multidecadal- to millennial-scale hydrologic variability on the South American Altiplano. Holocene 18:867–876
Fan Y, Duffy CJ, Oliver Jr DS (1997) Density-driven groundwater flow in closed desert basins: field investigations and numerical experiments. J Hydrol 196:139–184
Fritz SC, Baker PA, Lowenstein TK, Seltzer GO, Rigsby CA, Dwyer GS, Tapia PM, Arnold KK, Ku T-L, Luo S (2004) Hydrolic variation during the last 170,000 years in the southern hemisphere tropics of South America. Quat Res 61:95–104
Gasse F (2000) Hydrological changes in the African tropics since the last glacial maximum. Quat Sci Rev 19:189–211

Ghienne J-F, Schuster M, Bernard A, Duringer P, Brunet M (2002) The Holocene giant Lake Chad revealed by digital elevation models. Quat Int 87:81–85

Gilbert CM, Christensen MN, Al-Rawi Y, Lajoie KR (1968) Structural and volcanic history of Mono Basin, California-Nevada. Geol Soc Am Mem 116:275–331

Godfrey LV (2006) Focused groundwater flow and the accumulation of giant salt deposits in the Salar de Atacama. Geol Soc Am Abstr Programs 38:521

Goerner A, Jolie E, Gloaguen R (2009) Non-climatic growth of the saline Lake Beseka, Main Ethiopian Rift. J Arid Environ 73:287–295

Grosjean M (1994) Paleohydrology of the Laguna Lejia (north Chilean Altiplano) and climatic implications for late-glacial times. Palaeogeogr Palaeoclimatol Palaeoecol 109:89–100

Grosjean M, Geyh MA, Messerli B, Schotterer U (1995) Late-glacial and early Holocene lake sediments, ground-water formation and climate in the Atacama Altiplano 22–24°S. J Paleolimnol 14:241–252

Guyot JL, Roche MA, Noriega L, Calle H, Quintanilla J (1990)Salinities and sediment transport in the Bolivian highlands. J Hydrol 113:147–162

Hart WS, Quade J, Madsen DB, Kaufman DS, Oviatt CG (2004) The $^{87}Sr/^{86}Sr$ ratios of lacustrine carbonates and lake-level history of the Bonneville paleolake system. Geol Soc Am Bull 116:1107–1119

Hoke GD, Isacks BL, Jordan TE, Yu JS (2004) Groundwater-sapping origin for the giant quebradas of northern Chile. Geology 32:605–608

Holzbecher E (2005) Groundwater flow pattern in the vicinity of a salt lake. Hydrobiology 532:233–242

Isiorho SA, Matisoff G (1990) Groundwater recharge from Lake Chad. Limnol Oceanogr 35:931–938

Isiorho SA, Matisoff G, Wehn KS (1996) Seepage relationship between Lake Chad and the Chad aquifers. Ground Water 34:819–826

Jarsjo J, Destouni G (2004) Groundwater discharge into the Aral Sea after 1960. J Mar Syst 47:109–120

Johansson O, Aimbetov I, Jarsjo J (2009) Variation of groundwater salinity in the partially irrigated Amudarya River delta, Uzbekistan. J Mar Syst 76:287–295

Kafri U, Arad A (1972) Geochemistry of groundwaters in the Chad Basin. Rep. submitted to the F.A.O., Rome

Kebede E, Gebremariam Z, Ahlgren I (1994) The Ethiopian Rift Valley lakes: chemical characteristics of a salinity–alkalinity series. Hydrobiology 288:1–12

Kebede S, Travi Y, Asrat A, Alemayehu T, Ayenew T, Tessema Z (2008) Groundwater origin and flow along selected transects in Ethiopian rift volcanic aquifers. Hydrogeol J 16:55–73

Kebede S, Travi Y, Rozanski K (2009) The $\delta\,^{18}O$ and $\delta\,^{2}H$ enrichment of Ethiopian lakes. J Hydrol 365:173–182

Keir D, Stuart GW, Jackson A, Ayele A (2006) Local earthquake magnitude scale and seismicity rate for the Ethiopian Rift. Seism Soc Am Bull 96:2221–2230

Kohfahl C, Rodriguez M, Fenk C, Menz C, Benavente J, Hubberten H, Meyer H, Paul L, Knappe A, Lopez-Geta JA, Pekdeger A (2008) Characterising flow regime and interrelation between surface- water and ground-water in the Fuente de Piedra salt lake basin by means of stable isotopes, hydrogeochemical and hydraulic data. J Hydrol 351:170–187

Kowalewska A, Cohen AS (1998) Reconstruction of paleoenvironments of the Great Salt Lake Basin during the late Cenozoic. J Paleolimnol 20:381–407

Leblanc M, Favreau G, Maley J, Nazoumou Y, Leduc C, Stagnitti F, van Oevelen PJ, Delclaux F, Lemoalle J (2006) Reconstruction of Megalake Chad using Shuttle Radar Topographic Mission data. Palaeogeogr Palaeoclimatol Palaeoecol 239:16–27

Leblanc M, Favreau G, Tweed S, Leduc C, Razack M, Mofor L (2007) Remote sensing for groundwater modeling in large semiarid areas: lake Chad Basin, Africa. Hydrogeol J 15: 97–100

Lee K (1969) Infrared exploration for shoreline springs, a contribution to the hydrogeology of Mono Basin, California. PhD Thesis, Stanford University of California
Legesse D, Ayenew T (2006) Effect of improper water and land resource utilization on the central Main Ethiopian Rift lakes. Quat Int 148:8–18
Lowenstein TK, Risacher F (2009) Closed basin brine evolution and the influence of Ca-Cl inflow waters: Death Valley and Bristol Dry Lake California, Qaidam basin, China, and Salar de Atacama, Chile. Aquat Geochem 15:71–94
Magaritz M, Aravena R, Pena H, Suzuki O, Grilli A (1990) Source of ground water in the deserts of Northern Chile: evidence of deep circulation of ground water from the Andes. Ground Water 28:513–517
Manning AH, Solomon DK (2003) Using noble gases to investigate mountain-front recharge. J Hydrol 275:194–207
Micklin PP (1988) Desiccation of the Aral Sea: a water management disaster in the Soviet Union. Science 241:1170–1176
Micklin P (2007) The Aral Sea disaster. Annu Rev Earth Planet Sci 35:47–72
Micklin P, Aladin NV (2008) Reclaiming the Aral Sea. Sci Am 298:44–51
Oviatt CG, Miller DM, McGeehin JP, Zachary C, Mahan S (2005) The younger Dryas phase of Great Salt Lake, Utah, USA. Palaeogeogr Palaeoclimatol Palaeoecol 219:263–284
Pedone VA, Folk RL (1996) Formation of aragonite cement by nannobacteria in the Great Salt Lake, Utah. Geology 24:763–765
Peneva EL, Stanev EV, Stanychni SV, Salokhiddinov A, Stulina G (2004) The recent evolution of the Aral Sea level and water properties: analysis of satellite, gauge and hydrometeorological data. J Mar Syst 47:11–24
Rech JA, Quade J, Betancourt JL (2002) Late quaternary paleohydrology of the central Atacama Desert (lat 22–24°S), Chile. Geol Soc Am Bull 114:334–348
Reheis MC, Dixon TH (1996) Kinematics of the Eastern California shear zone: evidence for slip transfer from Owens and Saline Valley fault zones to Fish Lake Valley fault zone. Geology 24:339–342
Reheis MC, Stine S, Sarna-Wojcicki AM (2002) Drainage reversals in Mono Basin during the late Pliocene and Pleistocene. Geol Soc Am Bull 114:991–1006
Risacher F, Fritz B (1991) Quaternary geochemical evolution of the salars of Uyuni and Coipasa, Central Altiplano, Bolivia. Chem Geol 90:211–231
Risacher F, Fritz B (2000) Bromine geochemistry of the salar de Uyuni and deeper salt crusts, Central Altiplano, Bolivia. Chem Geol 167:373–392
Risacher F, Fritz B (2009) Origin of Salts and Brine Evolution of Bolivian and Chilean Salars. Aquat Geochem 15:123–157
Rogers DB, Dreiss SJ (1995) Saline groundwater in Mono Basin, California 1. Distribution. Water Resour Res 31:3131–3150
Rogers DB, Dreiss SJ (1995) Saline groundwater in Mono Basin, California 2. Long-term control of lake salinity by groundwater. Water Resour Res 31:3151–3169
Sagri M, Bartolini C, Billi P, Ferrari G, Benvenuti M, Carnicelli S, Barbano F (2008), Latest Pleistocene and Holocene river network evolution in the Ethiopian Lakes Region. Geomorphology 94:79–97
Seltzer GO, Baker P, Cross S, Dunbar R, Fritz S (1998) High-resolution seismic reflection profiles from Lake Titicaca, Peru-Bolivia: evidence for Holocene aridity in the tropical Andes. Geology 26:167–170
Shibuo Y, Jarsjo J, Destouni G (2007) Hydrological responses to climate change and irrigation in the Aral Sea drainage basin. Geophys Res Lett 34:L21406
Sieh K, Bursik M (1986) Most recent eruption of the Mono craters, eastern central California. J Geophys Res 91(B12):12539–12571
Small EE, Giorgi F, Cirbus Sloan L, Hostetler S (2001) The effects of desiccation and climatic change on the hydrology of the Aral Sea. J Climatol 14:300–322

Spencer RJ, Eugster HP, Jones BF (1985a) Geochemistry of Great Salt Lake, Utah II: Pleistocene–Holocene evolution. Geochim Cosmochim Acta 49:727–737

Spencer RJ, Eugster HP, Jones BF, Rettig SL (1985b) Geochemistry of Great Salt Lake, Utah I: hydrochemistry since 1850. Geochim Cosmochim Acta 49:739–747

Stephens DW (1990) Changes in lake levels, salinity and the biological community of Great Salt Lake (Utah, USA), 1847–1987. Hydrobiol 197:139–146

Stine S (1987) Mono Lake: the past 4,000 years. PhD Thesis, University of California, Berkeley

Stine S (1990) Past climate at Mono Lake. Nature 345:391

Stine S (1991) Geomorphic, geographic, and hydrographic basis for resolving the Mono Lake controversy. Environ Geol Water Sci 17:67–83

Sylvestre S, Servant M, Servant-Vildary S, Causse C, Fournier M, Ybert J-P (1999) Lake-level chronology on the southern Bolivian Altiplano (180–230S) during late glacial time and the early Holocene. Quat Res 51:54–66

Sylvestre F (2002) A high-resolution diatom reconstruction between 21,000 and 17,000 ^{14}C yr BP from the southern Bolivian Altiplano (18–23°S). J Paleolimnol 27:45–57

UNESCO (1969) Study of water resources in the Chad basin. Technical report, UNESCO/UNDP Special Fund Project Serial No. 1600/BMS.RD/SCE, 151 p

Valero-Garces BL, Delgado-Huertas A, Ratto N, Navas A, Edwards L (2000) Paleohydrology of Andean saline lakes from sedimentological and isotopic records, Northwestern Argentina. J Paleolimnol 24:343–359

Valero-Garces BL, Delgado-Huertas A, Navas A, Edwards L, Schwalb A, Ratto N (2003) Patterns of regional hydrological variability in central-southern Altiplano (18–26°) lakes during the last 500 years. Palaeogeogr Palaeoclimatol Palaeoecol 194:319–338

Vandervoort DS, Jordan TE, Zeitler PK, Alonso RN (1995) Chronology of internal drainage development and uplift, southern Puna plateau, Argentine central Andes. Geology 23:145–148

VanWormer JD, Ryall AS (1980) Sierra Nevada-Great Basin boundary zone: earthquake hazard related to structure, active tectonic processes, and anomalous patterns of earthquake occurrence. Seism Soc Am Bull 70:1557–1572

Wolfenden E, Ebinger C, Yirgu G, Deino A, Ayalew D (2004) Evolution of the northern Main Ethiopian rift: birth of a triple junction. Earth Planet Sci Lett 224:213–228

Yirgu G, Ebinger CG, Maguire PKH (2006) The Afar volcanic province within the East African Rift System: introduction. In: Yirgu G, Ebinger CG, Maguire PKH (eds) The Afar Volcanic Province within the East African Rift System. Geol Soc Lond Spec Publ 259:1–6

Chapter 10
Current Continental Base-Levels Below Sea Level and Distant from the Sea

10.1 General

The following continental base-levels are distant from the sea and have, assumingly, no connection to the current marine base-levels. They are defined herein as base-levels below sea level since their lowermost terminal portion is below sea level.

10.2 The Great Artesian Basin and Lake Eyre Basin, Australia

The Great Artesian Basin (hereafter GAB) (Fig. 8.1) is one of the largest and oldest groundwater basins in the world occupying a big portion of central Australia (Fig. 10.1).

The basin covers a surface area of some 1.711 million km^2 (Mudd 2000). The climate is semi-arid to arid, with low precipitation, which becomes higher, averaging 600 mm/y, in the elevated bounding mountains (Herczeg et al. 1991). The evaporation in the discussed basin is quite high varying between 2,500 and 3,000 mm/y.

The basin was formed in Mesozoic times, and it contains a sequence of confined Triassic to Cretaceous, detrital aquifers and alternating aquicludes, that attains a maximum thickness of 3,000 m. The configuration of the basin is of a huge synclinorium, whereby the aquifers are exposed to recharge mainly in the eastern margins of the Great Dividing Range in Queensland, and to a lesser degree in the western margins. This configuration results in a convergent groundwater flow to the Lake Eyre Basin which occupies a big portion of the GAB and covers a surface area of some 1.14 million km^2. Lake Eyre in the southwestern part of the basin serves as a continental endorheic terminal lake and base-level (Fig. 10.1) (Herczeg et al. 1991; Mazor 1995; Mudd 2000). The northern part of the GAB drains to the north to the Gulf of Carpentaria.

U. Kafri and Y. Yechieli, *Groundwater Base Level Changes and Adjoining Hydrological Systems*,
DOI 10.1007/978-3-642-13944-4_10,

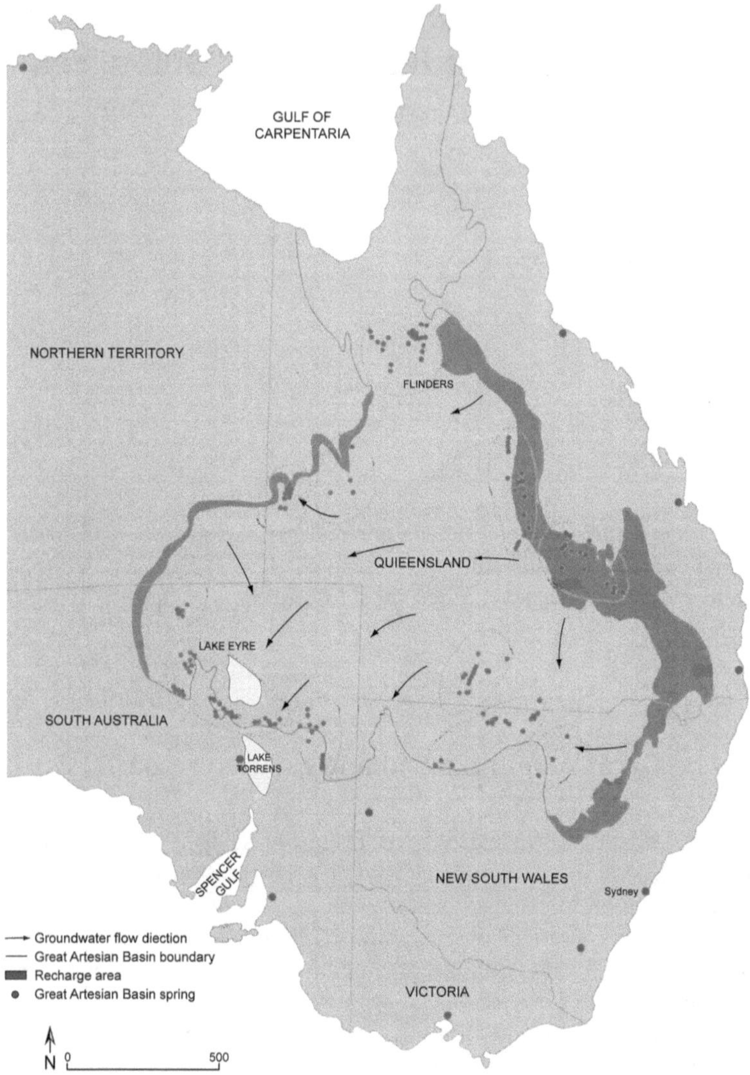

Fig. 10.1 A hydrological map of the Great Artesian Basin, Australia [after Mudd (2000)]

The aquifers of the GAB are confined and artesian, whereby their piezometric levels are above ground level over most of the GAB, excluding the margins of the basin.

In spite of being confined, the GAB is being discharged through vertical leakage to the regional water table and usually through hundreds of springs that emerge mostly close to the southern and southwestern margins of the basin (Figs. 10.1 and 10.2). Among these are the unique groundwater discharge features in the GAB of the artesian Mound Springs, that consist of carbonate deposited by the springs

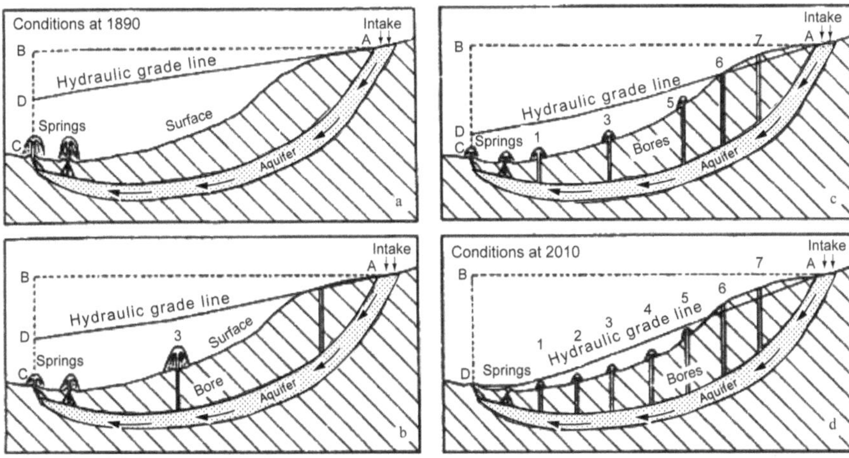

Fig. 10.2 Schematic hydrological sections in the Great Artesian Basin from the intake area to the discharge zone, showing the effect of groundwater exploitation on the decline of water heads and spring flows [after Mudd (2000)]

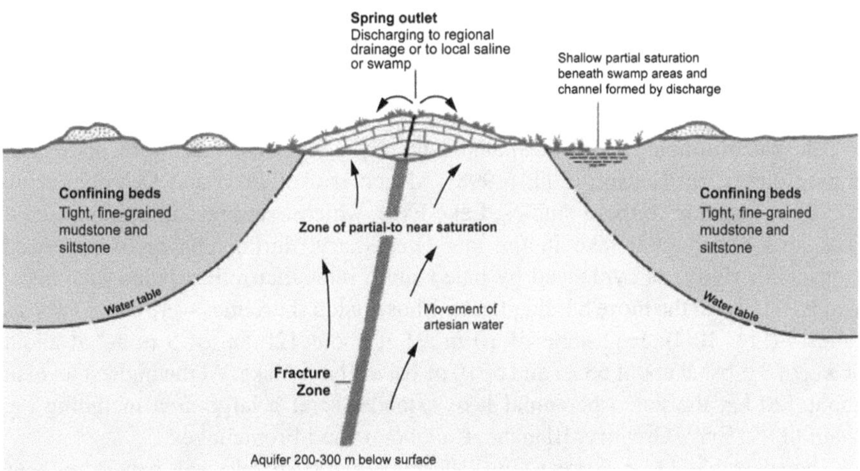

Fig. 10.3 The flow mechanism of a typical mound spring [after Mudd (2000)]

attaining, in places, an elevation of several meters (Fig. 10.3), as described in detail by Mudd (2000).

The age, as well as the residence and travel time of the groundwater of the GAB, from the intake areas to the discharge zones, were discussed in several articles. Mazor (1995) claimed that, due to the very slow groundwater flow in the basin, the aquifers are practically stagnant. Different from other studies, he attributed groundwater ages to be beyond the limits of the ^{36}Cl dating method, of one Ma. Namely, their ages are as old as 5–10 Ma or even more, based on the ^{4}He dating method.

Cresswell et al. (1999) described groundwater ages from the Amadeus Basin to be as old as 400 ka, and an average groundwater flow rate of 0.1 m/y, based on the ^{36}Cl dating method. Similarly, Collon et al. (2000) attributed to the GAB groundwater ages between 225 and 400 ka, based on the ^{81}Kr dating methods. In general, groundwater is relatively young, around several thousand years close to the intake areas and getting older along the flow path.

Groundwater salinity of the deeper Jurassic-early Cretaceous aquifer is between 500 and 1,000 mg/l, whereas the shallower Cretaceous aquifer already exhibits salinities of up to 10,000 mg/l (Mudd 2000). The geochemical characteristics of the GAB groundwater, as described by Herczeg et al. (1991), is as follows: The groundwater throughout the eastern and central parts of the basin is of the Na-HCO_3 type. Concentration of sodium and bicarbonate increases along the flow path toward the discharge zones in the west where the water changes to the Na-SO_4-Cl water type. The high Cl/Br weight ratio of these waters is indicative of evaporite dissolution.

Lake Eyre is the lowest terminal lake in the basin, at an elevation of 15 m bsl. The lake is practically an ephemeral huge playa or intermittent lake, formed by sporadic flooding, as happened several times during the last decades. Water salinity changes drastically in accordance to flooding and desiccation events. As an example, during one year, between 1984 and 1985 water salinity changed from 25 to 270 g/l (Williams and Kokkinn 1988). Evaporite beds and crusts are formed when water salinity reaches the saturation stage during the desiccation and dry time spans.

The paleolimnology and the paleo Lake Eyre level changes were discussed, among others, by Nanson et al. (1998), Magee et al. (2004) and DeVogel et al. (2004). According to these studies, Lake Eyre, which is at present an intermittent lake, was a perennial lake in the late Quaternary, during phases of increased monsoon activity, as evidenced by paleo shore lines that reflect paleo lake high-stands, related to the more humid phases. Those paleo shorelines were dated to be as follows (Fig. 10.4): Highstand of 10 m asl at about 120 ka, of 5 m asl at about 86 ka, of 3.5 bsl at about 63 ka and of 10 m bsl at about 40 ka. At the highest level at about 120 ka, the paleo perennial lake extended over a large area including the chain of the Eyre, Gregory, Blanche, Callabonna and Frome lakes.

The endorheic Lake Torrens (Fig. 10.1) is a 5,700 km^2 dry salt pan, a few tens of meter above sea level, south of Lake Eyre. Both lakes are separated by a topographic divide with a lowest elevation at 92 m asl indicating that paleo Lake Eyre could have never over-flown southward to Lake Torrens (DeVogel et al. 2004). The chloride content of the groundwater increases towards Lake Torrens and its brine enrichment occurs in the playa by evaporation from the capillary zone (Schmid 1988). It is important to note that Lake Eyre and Lake Torrens are aligned along a rift or a trough, which extends from the Spencer Gulf in the south, and is bounded to the east by the Flinders Ranges. Several studies showed that the area has been tectonically active, evidenced by subsidence of the rift, by young displacements, by current seismicity and by anomalous heat flow (Neumann et al.

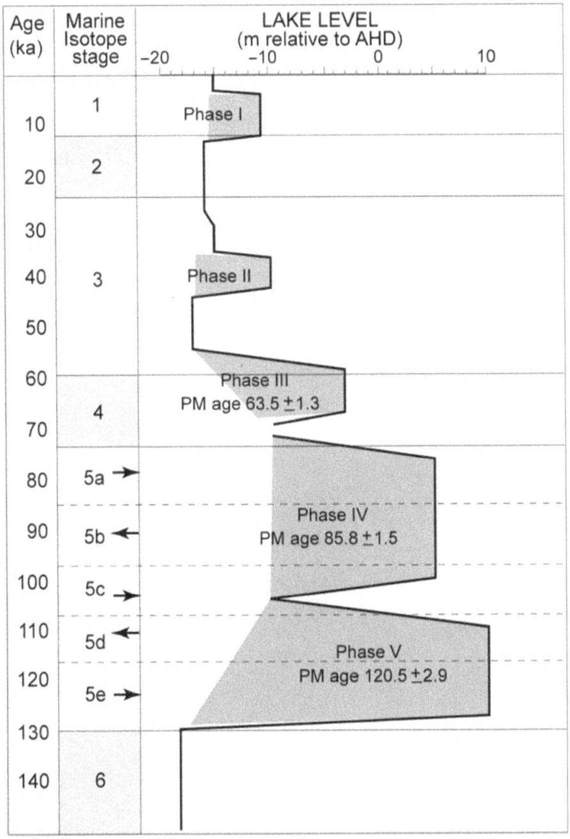

Fig. 10.4 Lake Eyre level curve for the past 150 ka [modified after Magee et al. (2004)]. PM ages denote pooled mean ages

2000; Sandiford 2003; Celerier et al. 2005; Quigley et al. 2006, 2007; Sandiford and Quigley 2009).

The Lake Eyre endorheic basin exhibits a combination of a continental base-level, only somewhat below sea level, located along a tectonically active trough which extends to the sea, and separated from the latter by an assumed rather low groundwater divide in between. Coupled with the existence of a geothermal anomaly it can be speculated that a subsurface seawater encroachment to the lower base-level may occur along the active faults. This process could be problematic in this case because of the long distance, and the relative, small elevation difference between Lake Eyre and the sea of only 15 m. The proposed additional mechanism of salination herein, by encroaching seawater, might be supported by the convecting seawater mechanism, tied to the existing geothermal anomaly, that was proposed for the Salton Trough by Barragan et al. (2001). The specific condition which favor salination due to seawater intrusion from the sea into low lying base level of the Salton Trough are elaborated in Sect. 11.5.

10.3 The Death Valley Basin

The Death Valley (Fig. 8.1) is a endorheic closed basin, at coordinates 36°N 117°W, in California, USA, attaining its lowest elevation at 86 m bsl. It is a north-south directed elongated valley, between the Amargosa Range (Black and Funeral Mountains) in the east and the Panamint range in the west (Fig. 10.5).

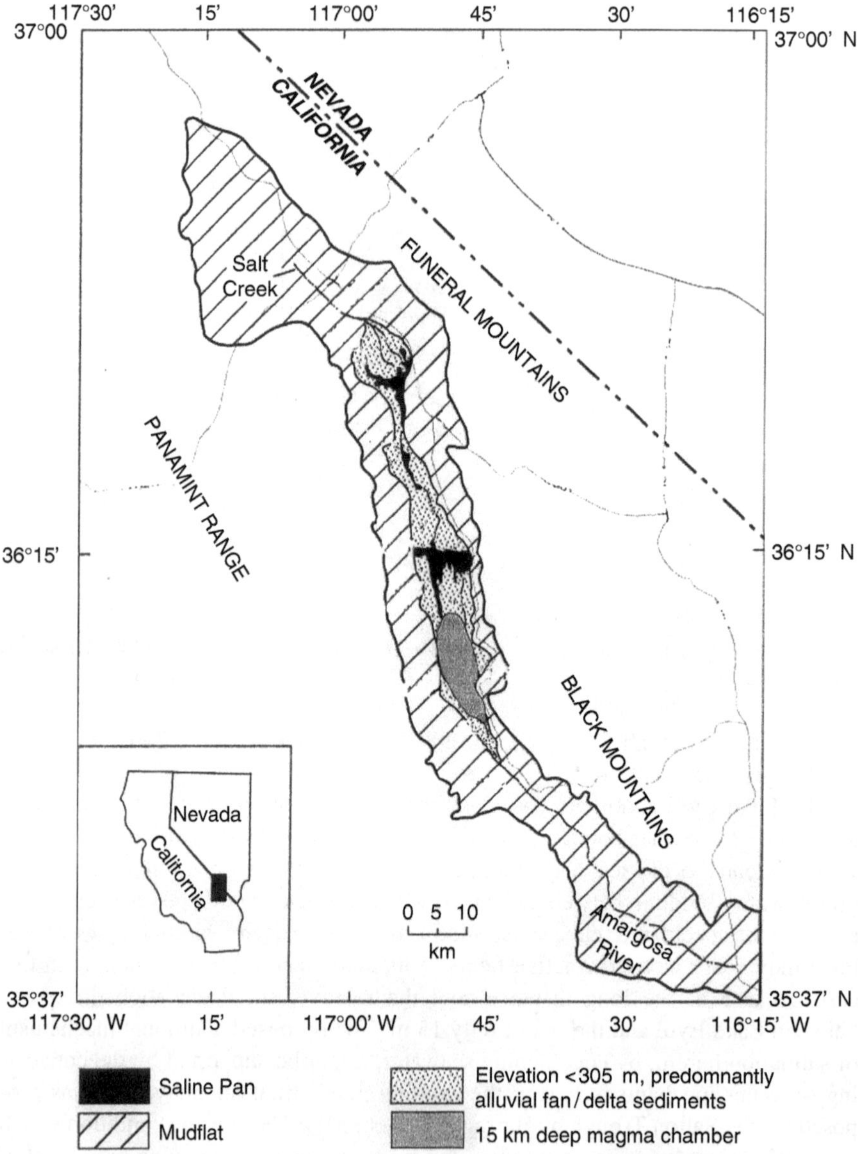

Fig. 10.5 Map of the Death Valley, California [after Lowenstein and Risacher (2009)]

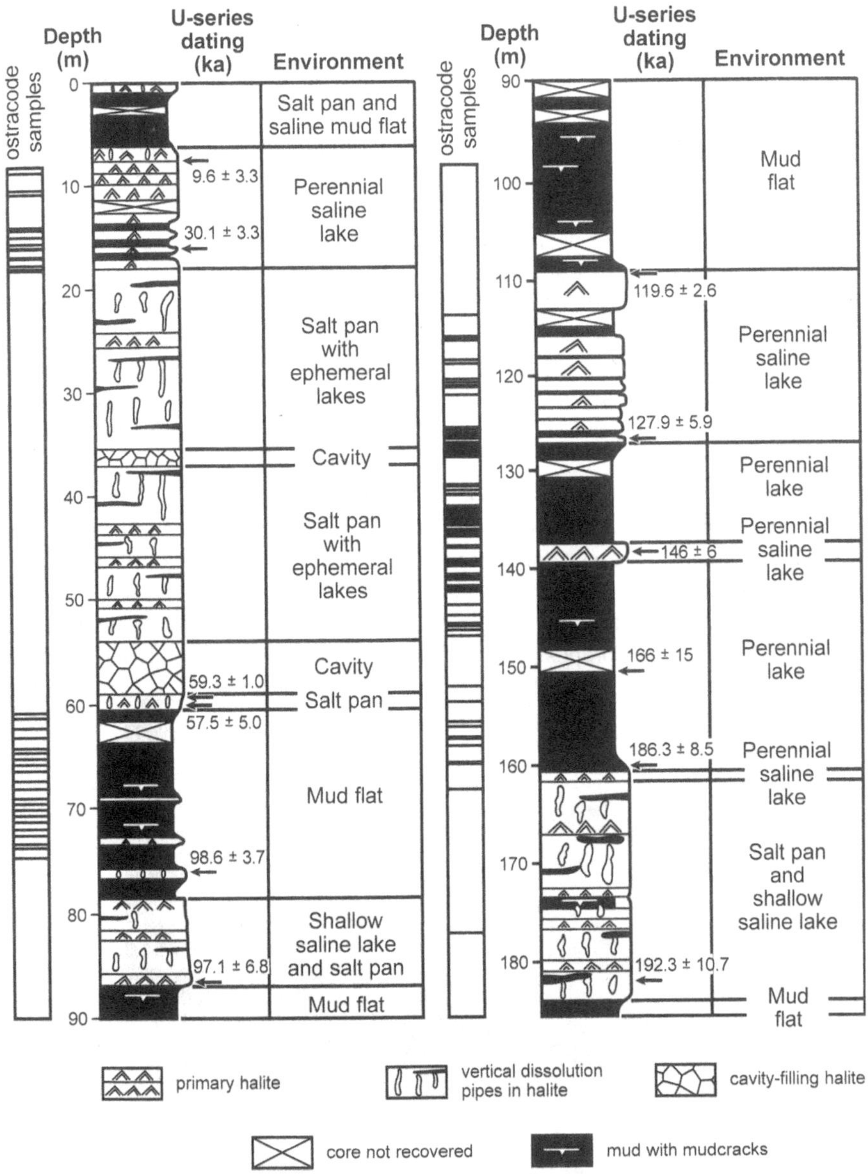

Fig. 10.6 A core columnar section in the Death Valley exhibiting the paleohydrological changes during the last 200 ka [after Lowenstein et al. (1999)]

Structurally speaking, the basin is part of the Basin Range province, formed as a pull apart basin and associated with strike slip and normal faults, which are still active (Lowenstein and Risacher 2009). The climate is arid, typified by an average precipitation of no more than 50 mm/y, and high temperatures and evaporation. The bottom of the valley consists of mud flats, salt pans and crusts (Fig. 10.5), as described by Crowley and Hook (1996) and Lowenstein et al. (1999).

The paleoclimate and paleohydrology of the basin were studied by several authors, among which are: Li et al. (1996, 1997), Ku et al. (1998), Lowenstein et al. (1999), Nelson et al. (2001), Forester et al. (2005) and Miner et al. (2007). The indicators that were used to reconstruct the paleohydrological regime of the basin were sedimentology, water salinity, stable and radiogenic isotopes of water, paleo spring deposits and ostracods. According to the above studies, the basin exhibits a closed basin characteristics in the last 200 ka, subjected to climatic changes and alternating hydrological regimes evidenced by the following: Dry periods, similar to the present one, are evidenced by mud flats, shallow saline lakes and salt pans which prevailed between 110 and 60 ka, and between 10 ka and the present. More humid periods are evidenced by perennial fresher lakes and perennial saline lakes, such as the occupied paleo Lake Manley between 185 and 128 ka and between 35 and 10 ka (Fig. 10.6).

The present day inflows to the Death Valley endorheic basin include rivers inflow from north and south and convergent groundwater and spring water flows that are recharged in the surrounding mountains intake areas (Fig. 10.7) (Bakker et al. 1999; Lowenstein et al. 1999; Anderson et al. 2006) .This pattern is supported mainly by the minimal direct recharge, related to the present arid climate, and the depleted stable isotope compositions of the groundwater, related to higher recharge altitudes (Larsen et al. 2001) However, Anderson et al. (2006) disagree with the

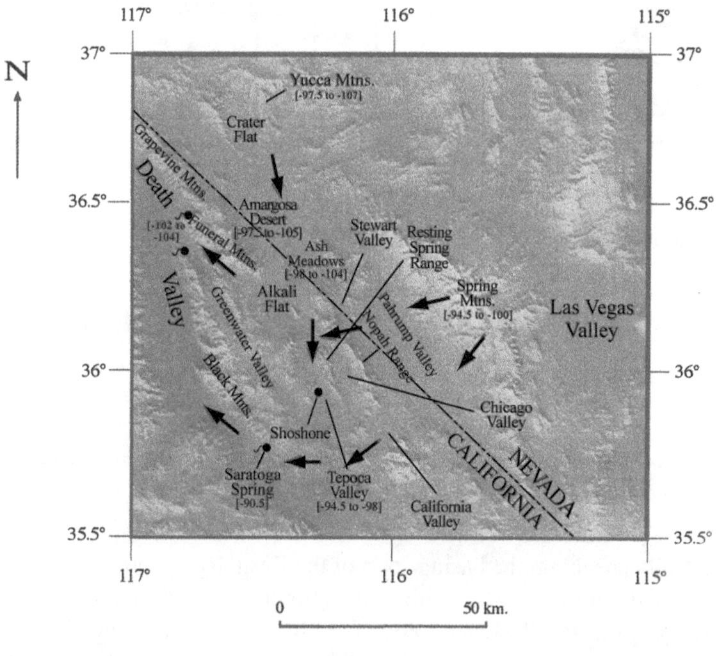

Fig. 10.7 Regional groundwater flow paths in the Death Valley area [after Nelson et al. (2001)]

conventional model of inter-basinal flow from distant parts of the basin, claiming that most of the recharge is local.

Higher salinities of groundwater are reached in the basin through interaction with saline lakes deposits (Larsen et al. 2001). An additional suggested saline contribution is the inflow to the basin of thermal waters from depth (Larsen et al. 2001), or more specific, hydrothermal Ca-Cl brines which emerge along active faults associated with an existing magma chamber at depth. This mechanism resembles the case of the Andean Altiplano and the Qaidam base-level in China, as described by Lowenstein and Risacher (2009) (see also Sects. 9.4 and 10.4).

The Death Valley base-level resembles, regarding its features and the hydrological regime, other endorheic continental base levels, described in this chapter.

10.4 Central Asia Closed Endorheic Basins

The vast region of central Asia (Fig. 8.1), which includes northwest China and Inner Mongolia, occupies hundreds of closed inland terminal river basins and lakes at elevations between 154 m bsl and thousands of meter above sea level. The basins are a few thousands kilometer distant from the nearest marine base level. These basins serve as base-levels to both surface and groundwater flows, occupying fresh as well as saline lakes, dry deserts and salt pans. This region is included technically, herein, under the category of base-levels below sea level due to its lowermost elevation, despite the higher elevations of most of its endorheic basins as described below.

The basins are, in general, inter-mountain basins situated structurally in between long mountain ranges such as the Altay, Tianshan, Qilian and Kulun mountains (Fig. 10.8). The entire region was subjected during its geological history to tectonic movements which culminated, induced by the Indian–Eurasian plates collision, since late Pliocene – early Pleistocene (Tapponnier and Molnar 1977). The mountain ranges were uplifted at a higher rate and the basins in between subsided, accumulating vast amounts of detrital basin fill sediments derived from the mountains, as well as lacustrine deposits (Edmunds et al. 2006; Zhu et al. 2007).

The paleo climatic and paleo hydrological regime of the region was reconstructed in several studies, based on paleo lakes shore lines, stable and radiogenic isotope compositions of the groundwater, sedimentological characteristics and pollen assemblages. According to these studies, the region was subjected to dry and humid climatic variations and thus to resultant different hydrological regimes since the late Pleistocene. The basins of northwestern China exhibit a warm and a more humid climate than the present one, during 40–30 ka BP (Yang et al. 2004). The region was also subjected to frequent climatic variations between humid and dry periods during the entire Holocene, between 10.7 and 1.5 ka BP (Hartmann and Wuenemann 2009), as is also evidenced by pollen assemblages (Feng et al. 2006).

Among the known closed basins of the region, those that are referred to below, are: The Tarim, Heihe, Minqin and Qaidam basins, all of which are, regarding their

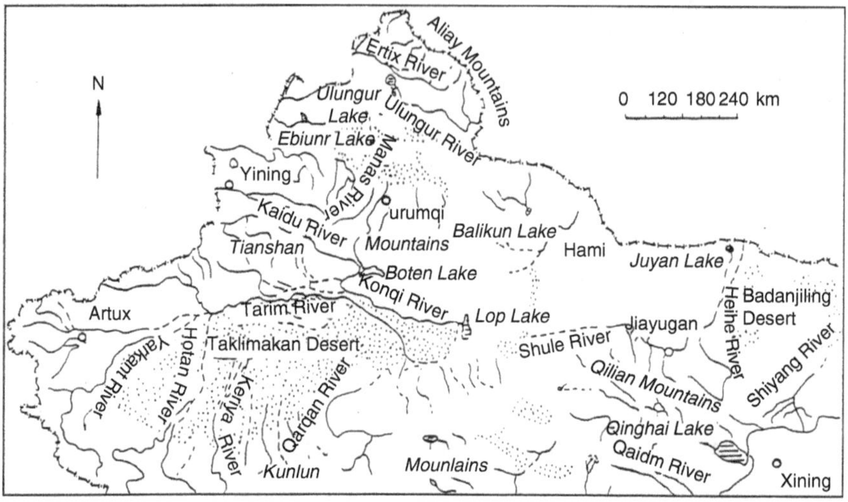

Fig. 10.8 Map of arid zones and stream systems in northwest China [after Wang and Cheng (2000)]

elevation, above sea level and the Turfan (Turpan) Basin which is considerably below sea level at its lower reaches. All these basins are characterized by an arid climate.

The Tarim Basin, is situated between the Tianshan Mountains in the north and the Kunlun Mountains in the south. Its terminal discharge zone is the Lop Nor Lake at coordinates 40°N 90°E and at the elevation of 768 m asl. The basin is drained by the largest inland river in China. The lower reaches of the basin and the lake are in the process of desiccation related to climate change and groundwater abstraction accompanied by water quality degradation. Groundwater salinity in the lower reaches of the basin rose to the level of 13 g/l (Feng et al. 2001, 2005).

The Heihe Basin extends from the Qilian Range in the south to its terminal Ejina Basin and Juyan lakes, at coordinates 41°N 100°E, in the lower reaches of the Heihe River. The Ejina Basin serves as a terminal discharge zone to both surface and groundwater flows. The convergent groundwater flow towards the Juyan lake in the north is displayed in the water table contour map of the basin (Fig. 10.9) (Wen et al. 2005). Groundwater salinity attained levels that exceed 5 g/l, in some places in the basin (Zhu et al. 2008). In the terminal northern part of the basin, the salinity is probably significantly higher as evidenced by the abundant salt crusts, as a result of super-saturation through evaporation (Wen et al. 2005).

The Minqin Basin, at coordinates 38°N 103°E, is situated at the lower reaches of the Shiyang River Basin, which extends from the Qilian Range in the south to the Minqin Basin in the north. The basin fill in the north is rich in evaporites such as halite and gypsum. The salinity of groundwater in the basin attained a level that exceeds 5 g/l. The basin was subjected to water table drop and water salination following severe over-exploitation (Zhu et al. 2007).

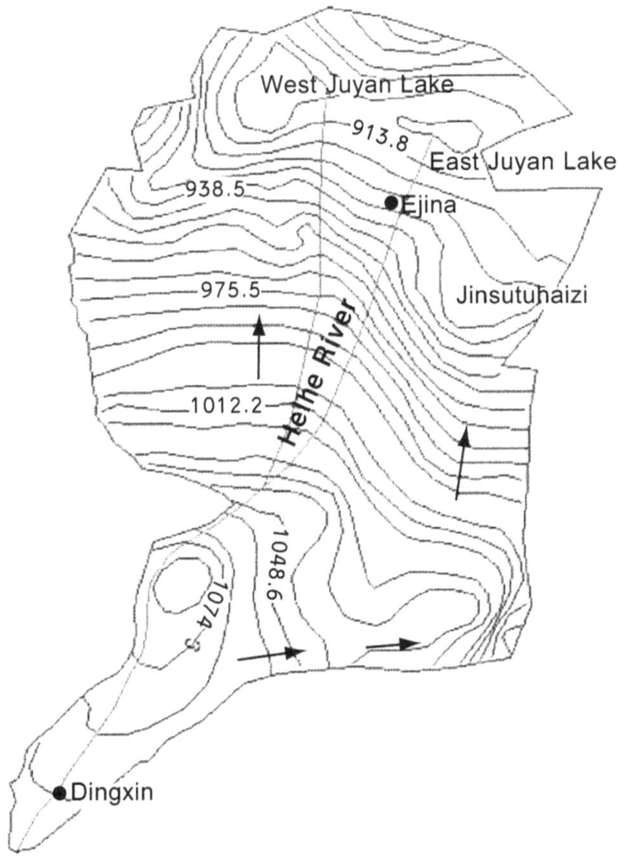

Fig. 10.9 Groundwater table map in the Ejina Basin exhibiting convergent flow toward the Juyan Lake system in the north [after Wen et al. (2005)]

The Qaidam Basin, at coordinates 37°N 94°E, is a terminal closed basin between the Kunlun Mountains in the south and the Qilian Mountains in the north. The center of the basin contains large areas of salt pans and saline lakes with the largest Qaharan Salt Lake plain. The Qaharan salt plain contains concentrated brines some of which are Ca-Cl brines (Lowenstein and Risacher 2009). Inflows to the center of the basin are from rivers that originate in the surrounding mountains, from converging groundwater flow from the mountain aquifers and from emerging hydrothermal Ca-Cl brines from depths, as found in other similar (Andean Altiplano, Death Valley) basins (Lowenstein and Risacher 2009) (see also Sects. 9.4 and 10.3).

The Turpan (Turfan) Basin is a terminal endorheic basin, at coordinates 42°N 90°E. The basin is structurally a fault bounded trough between the Tianshan Mountains in the north and the Quoltag Mountains in the south. The lowest point in the basin is the saline Lake Ayding at 154 m bsl that serves as the terminal base-level to surface and groundwater flows. The salinity of the lake is extremely high, exceeding 200 g/l. The inflows to the basin is only via rivers and groundwater that are fed by rainfall and snow melt and recharge that originate on the Tianshan Mountains in the north (Wei and Gasse 1999).

The present climate is generally arid in the lower parts of the basins and is more humid in the surrounding mountain ranges. As a result, the basins' aquifers are subjected to almost no direct recharge from rainfall in their lower reaches, but rather from the rivers that originate on the mountain ranges fed by rainfall as well as of snow and glacier melt and by lateral groundwater flow. Owing to the configuration of mountain ranges and closed basins, the rivers form a centripetal stream systems draining into the inland lakes or disappearing in the depression (Wang and Cheng 2000).

The basins' aquifers are recharged not only by groundwater convergent flows from the mountains aquifers (Edmunds et al. 2006) but also by the interconnected surface-groundwater systems as shown schematically in Fig. 10.10 (Wang and Cheng 2000; Ji et al. 2006).

A general down gradient geochemical evolution and zonation exist from the intake areas to the discharge zones, from fresh, mostly bicarbonate water types to farther more saline waters of bicarbonate–sulfate, sulfate–chloride and chloride water types (Ji et al. 2006). The terminal lower reaches of the basins occupy either brackish waters or hypersaline brines, attaining a salinity of up to 350 g/l in the Xingjiang lakes and up to almost 500 g/l of their adjoining interstitial waters (Zheng 1987). Additional potential salinity sources are emerging Ca-Cl brines, as described before from the Qaidam Basin by Lowenstein and Risacher (2009).

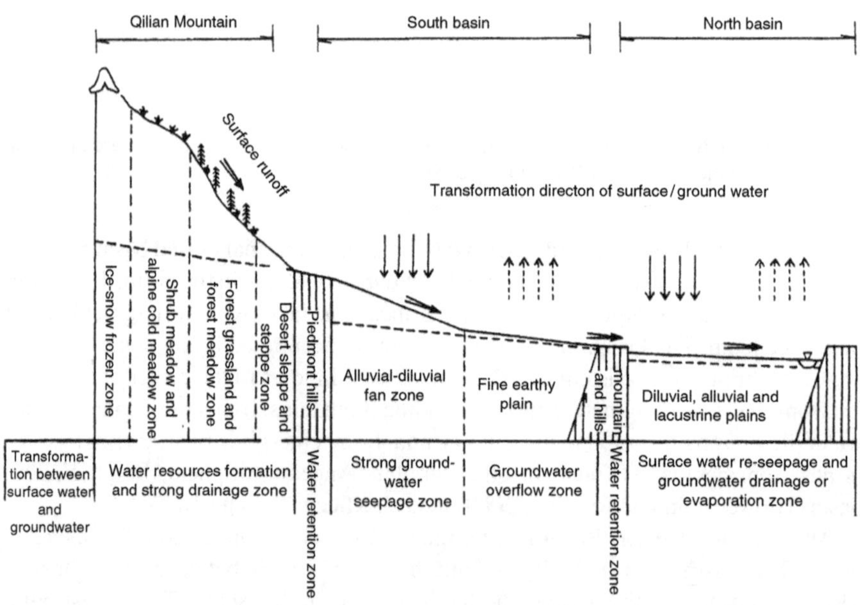

Fig. 10.10 A sketch of the hydrological regime of the different zones in the inland river basins of northwest China [after Wang and Cheng (2000)]

The geochemical evolution of groundwater in the discussed basins, along the flow path from the mountains intake areas to the terminal discharge zones, as well as the groundwater ages, were discussed in several studies (e.g., Zheng 1987; Feng et al. 2001, 2005; Wen et al. 2005; Edmunds et al. 2006; Ji et al. 2006; Chen et al. 2006; Zhu et al. 2007; Zhu et al. 2008; Lowenstein and Risacher 2009; and Su et al. 2009).

The hydrological characteristics, based on the above, that prevail in the discussed basins are as follows: The hydrological system consists, basically, of the mountain aquifers that are phreatic in the intake areas, and turn to be confined down-gradient within the lower parts of the basins. On top of that, a shallower phreatic aquifer system exists adjacent to the base levels.

Regarding stable isotope composition and groundwater age, it is noticed that the confined aquifers have a depleted stable isotope composition as compared to that of present day precipitation. This is in agreement with the tritium free content and the older ^{14}C ages of the groundwater, which is indicative of mostly paleo recharge in more humid periods. The shallow phreatic aquifers, on the other hand, contain tritium and are isotopically heavier than that of the present day precipitation, as a result of continuous evaporation along the flow path.

Regarding salinity and geochemical evolution of the groundwater, it is noticed that the shallow phreatic aquifers are more subjected to increase of salinity, as compared to the confined aquifers. The processes that are responsible for the salinity rise down-gradient are evaporation, water–rock interaction, dissolution of evaporites, over-exploitation, and to anthropogenic pollution.

10.5 The Caspian Sea-Kara Bogaz Gol Basin

The Caspian Sea basin (Fig. 8.1) is the largest closed terminal water body on earth, with a surface area of ca 380,000 km^2, located at coordinates 40°N 51°E (Fig. 10.11). Its present elevation is around 28 m bsl and it is fed mainly by the inflows of the Volga River which contributes over 80% of its water input, and the rest is contributed by the Kura and Ural rivers. Regarding the history of the basin, it started as part of the initial giant regional Sarmatian Sea in the Miocene around 17 Ma, which was later on, around 6 Ma, separated into the Caspian Sea, the Black Sea and Panonian basins. The Caspian Sea was later separated from the Black Sea at around 3 Ma (Klige and Myagkov 1992). Different from other authors, it was claimed by Dumont (1998) that the Caspian Sea was never connected to the Tethys Sea and should thus be regarded and termed as a lake.

In sub-recent historical times, the sea level fluctuated by some 6 m, but during its older geological history, fluctuations were over 200 m and the sea remained below sea level since the last glacial period (Dumont 1998). The level of the Caspian Sea fluctuated throughout the Pleistocene (Dumont 1998; Zubakov 2001) and throughout the last 2,500 years, but did not decline below 25 m bsl (Rychagov 1997). A dramatic water level decline of some 3 m was recorded during the 1930s (Giralt et al. 2003).

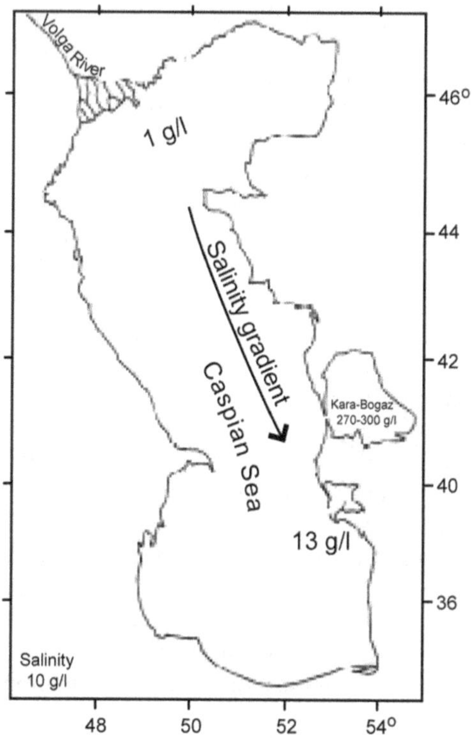

Fig. 10.11 Map of the Caspian Sea-Kara Bogaz system [modified after Dumont (1998)]. Also shown is the water salinity at various parts of the system

The hydrology and water chemistry of the present day Caspian Sea was studied, among others, by Dumont (1998), Clauer et al. (2000, 2009), Giralt et al. (2003) and Tuzhilkin et al. (2005). The Caspian Sea, at present, is an elongated water body, subdivided into three sub-basins (Giralt et al. 2003). It is shallow in the north, opposite the inlet and the delta of the Volga River and deeper in its central and southern portions. The vast amount of inflow from the Volga River to the shallow northern part of the sea, coupled with a more humid and cooler climate, as compared to the more arid southern parts of the sea, form a water salinity gradient to the south (Fig. 10.11). At the northwest, adjacent to the delta of the Volga River salinities are very low, whereas in the southeast salinity rises to 13 g/l, close to a third of ocean waters (Dumont 1998). The sea displays a significant multi-annual and spatial hydrochemical variability as reflected by the total water salinity of 7.2, 13 and 13.1 g/kg of the northern, central and southern sub-basins, respectively (Tuzhilkin et al. 2005).

The salinity rise toward the south occurs mostly through evaporation, but it was suggested, based on certain extra ion enrichment and specific isotope ratios (i.e., $^{87}Sr/^{86}Sr$), that an additional subterranean emerging hydrothermal water contribution also exists. This water leaches, while emerging, sedimentary evaporites, and is partly responsible for the specific water chemistry of the sea water. It also contributes to the rise of the sea level since about 1978 (Clauer et al. 2000, 2009). The main evaporite deposits are sulfates, as expected considering the water salinity.

The Kara Bogaz Gol Lake (KBG), east of the Caspian Sea, at coordinates 41°N 54°E (Fig. 10.11), is a shallow lagoon, a few meter deep with a surface area of some 18,000 km^2, located in an area of extremely arid climate. The hydrology of the KGB was discussed, among others, by Giralt et al. (2003), Leroy et al. (2006) and Kosarev et al. (2009). The KBG water level is several tens of centimeter to a few meter lower than that of the Caspian Sea and both are connected by a narrow strait, 110–300 m wide and 10–12 km long, through where the Caspian Sea overflows to the KBG. Following a considerable drop of the Caspian Sea level since the late 1970s, the straits were dammed, in March 1980, in order to stop this drop, and thus the inflow from the Caspian Sea to the KBG ceased. Following the closure of the straits, a quick desiccation of the KBG began, until it turned to a dry salt pan at the end of 1984. In 1992 the inflow from the Caspian Sea was renewed following the destruction of the dam.

The water salinity evolution of the KBG is controlled by the amount of Caspian Sea water inflow. Until the 1930s, the salinity was dominated by the vast influx of Caspian Sea waters of relatively moderate salinity and, thus, mostly sulfates (gypsum) were deposited at the bottom of the lagoon. Already in 1939 the water salinity increased, dominated mostly by a NaCl brine and halite started to be deposited since then. Following the drop of the KBG level in the 1970s, the water salinity rose to 27–30% and following the closure and disconnection from the Caspian Sea the brine salinity attained a concentration of up to 38%, accompanied by salt deposition of halite, epsomite, bischofite and carnallite.

The Caspian Sea is conventionally regarded everywhere as a terminal base-level. This statement is true when the entire basin, including the KBG is concerned. It is suggested herein that the Caspian Sea, proper, is in fact not naturally terminal by definition, since it overflows to the KBG and, thus, can be determined as a flow-through basin. As a result of this setup, the water salinity of the Caspian Sea remains intermediate, able to deposit sulfates only, such as gypsum, and not reaching salinities which are supersaturated with regard to halite. However, the KBG lagoon, downstream, is indeed the terminal part of the basin whereby salinities due to evaporation exceed the saturation threshold for halite and post-halite salts.

The Karagiye Depression, at coordinates 43°N 51°E, east of the Caspian Sea is a depression attaining an elevation as low as 132 m bsl. The entire area is described as saline lowlands. In spite of the scarce information, one can assume that this depression can potentially serve as an additional base-level to eastward density-driven brines from the Caspian Basin. Another depression, more to the east, at coordinates 41°N 58°E, namely the Akdzhakaya Depression, attains an elevation of 81 m bsl and is also described as a saline lowland. One can speculate that this depression functions, regarding salination processes, similar to the Karagiye Depression.

References

Anderson K, Nelson S, Mayo A, Tingey D (2006) Interbasin flow revisited: the contribution of local recharge to high-discharge springs, Death Valley, CA. J Hydrol 323:276–302

Bakker M, Anderson EI, Olsthoorn TN, Strack ODL (1999) Regional groundwater modeling of the Yucca Mountains site using analytic elements. J Hydrol 226:167–178

Barragan RRM, Birkle P, Portugal ME, Arellano GVM, Alvarez R (2001) Geochemical survey of medium temperature geothermal resources from the Baja California Peninsula and Sonora, Mexico. J Volcanol Geotherm Res 110:101–119

Celerier J, Sandiford M, Hansen DL, Quigly M (2005) Modes of active intraplate deformation, flinders ranges, Australia. Tectonics 24:TC6006:1–17

Chen Z, Nie Z, Zhang G (2006) Environmental isotopic study on the recharge and residence time of groundwater in the Heihe River Basin, northwestern China. Hydrogeol J 14:1635–1651

Clauer N, Chaudhuri S, Toulkeridis T, Blanc G (2000) Fluctuations of Caspian Sea level: beyond climatic variations?. Geology 28:1015–1018

Clauer N, Pierret M-C, Chaudhuri S (2009) Role of subsurface brines in salt balance: the case study of the Caspian Sea and Kara Bogaz Bay. Aquat Geochem 15:237–261

Collon P, Kutschera W, Loosli HH, Lehmann BE, Purtschert R, Love A, Sampson L, Anthony D, Cole D, Davids B, Morrissey DJ, Sherrill BM, Steiner M, Pardo RC, Paul M (2000) ^{81}Kr in the Great Artesian Basin, Australia: a new method for dating very old groundwater. Earth Planet Sci Lett 182:103–113

Cresswell RG, Jacobson G, Wischusen J, Fifield LK (1999) Ancient groundwaters in the Amadeus Basin, Central Australia: evidence from the radio-isotope ^{36}Cl. J Hydrol 223:212–220

Crowley JK, Hook SJ (1996) Mapping playa evaporite minerals and associated sediments in Death Valley, California, with multispectral thermal infrared images. J Geophys Res 101:643–660

DeVogel SB, Magee JW, Manley WF, Miller GH (2004) A GIS-based reconstruction of late quaternary paleohydrology: Lake Eyre, arid central Australia. Palaeogeogr Palaeoclimatol Palaeoecol 204:1–13

Dumont HJ (1998) The Caspian Lake: history, biota, structure, and function. Limnol Oceanogr 43:44–52

Edmunds WM, Ma J, Aeschbach-Hertig W, Kipfer R, Darbyshire DPF (2006) Groundwater recharge history and hydrogeochemical evolution in the Minqin Basin, North West China. Appl Geochem 21:2148–2170

Feng Q, Endo KN, Cheng GD (2001) Towards sustainable development of the environmentally degraded arid rivers of China – a case study from Tarim River. Environ Geol 41:229–238

Feng Q, Liu W, Si J, Su Y, Zhang Y, Cang Z, Xi H (2005) Environmental effects of water resource development and use in the Tarim River basin of northwestern China. Environ Geol 48:202–210

Feng X, Yan S, Ni J, Kong Z, Yang Z (2006) Environmental changes and lake level fluctuation recorded by lakes on the plain in northern Xinjiang during the late Holocene. Chin Sci Bull 51 Suppl 1:60–67

Forester MR, Lowenstein TK, Spencer RJ (2005) An ostracod based paleolimnologic and paleohydrologic history of Death Valley: 200 to 0 ka. Geol Soc Am Bull 117:1379–1386

Giralt S, Julia R, Leroy S, Gasse F (2003) Cyclic water level oscillations of the KaraBogazGol-Caspian Sea system. Earth Planet Sci Lett 212:225–239

Hartmann K, Wuenemann B (2009) Hydrological changes and Holocene climate variations in NW China, inferred from lake sediments of Juyanze palaeolake by factor analyses. Quat Int 194:28–44

Herczeg AL, Torgersen T, Chivas AR, Habermehl MA (1991) Geochemistry of ground waters from Great Artesian Basin, Australia. J Hydrol 126:225–245

Ji X, Kang E, Chen R, Zhao W, Zhang Z, Jin B (2006) The impact of the development of water resources on environment in arid inland river basins of Hexi region, Northwestern China. Environ Geol 50:793–801

Klige RK, Myagkov MS (1992) Changes in the water regime of the Caspian Sea. Geo J 27: 299–307

Kosarev AN, Kostianoy AG, Zonn IS (2009) Kara-Bogaz-Gol Bay: physical and chemical evolution. Aquat Geochem 15:223–236

Ku T-L, Luo S, Lowenstein TK, Li J, Spencer RJ (1998) U-series chronology of lacustrine deposits in Death Valley, California. Quat Res 50:261–275

Larsen D, Swihart GH, Xiao Y (2001) Hydrochemistry and isotope composition of springs in the Tecopa basin, southeastern California, USA. Chem Geol 179:17–35

Leroy SAG, Marret F, Giralt S, Bulatov SA (2006) Natural and anthropogenic rapid changes in the Kara-Bogaz Gol over the last two centuries reconstructed from palynological analyses and a comparison to instrumental records. Quat Int 150:52–70

Li J, Lowenstein TK, Blackburn IR (1997) Response of evaporate mineralogy to inflow water sources and climate during the past 100 k.y. in Death Valley, California. Geol Soc Am Bull 109:1361–1371

Li J, Lowenstein TK, Brown CB, Ku T-L, Luo S (1996) A 100 ka record of water tables and paleoclimates from salt cores, Death Valley, California. Palaeogeogr Palaeoclimatol Palaeoecol 123:179–203

Lowenstein KT, Li J, Brown C, Roberts SM, Ku T-L, Luo S, Yang W (1999) 200 k.y. paleoclimate record from Death Valley salt core. Geology 27:3–6

Lowenstein TK, Risacher F (2009) Closed basin brine evolution and the influence of Ca-Cl inflow waters: Death Valley and Bristol Dry Lake California, Qaidam basin, China, and Salar de Atacama, Chile. Aquat Geochem 15:71–94

Magee JW, Miller GH, Spooner NA, Questiaux D (2004) Continuous 150 k.y. monsoon record from Lake Eyre, Australia: insolation-forcing implications and unexpected Holocene failure. Geology 32:885–888

Mazor E (1995) Stagnant aquifer concept Part 1. Large-scale artesian systems-Great Artesian Basin, Australia. J Hydrol 173:219–240

Miner RE, Nelson ST, Tingey DG, Murrell MT (2007) Using fossil spring deposits in the Death Valley region, USA to evaluate palaeoflowpaths. J Quat Sci 22:373–386

Mudd GM (2000) Mound springs of the Great Artesian Basin in South Australia: a case study from Olympic Dam. Environ Geol 39:463–476

Nanson GC, Callen RA, Price DM (1998) Hydroclimatic interpretation of quaternary shorelines on South Australian playas. Palaeogeogr Palaeoclimatol Palaeoecol 144:281–305

Nelson ST, Karlsson HR, Paces JB, Tingey DG, Ward S, Peters MT (2001) Paleohydrologic record of spring deposits in and around Pleistocene pluvial Lake Tecopa, southeastern California. Geol Soc Am Bull 113:659–670

Neumann N, Sandiford M, Foden J (2000) Regional geochemistry and continental heat flow: implications for the origin of the South Australian heat flow anomaly. Earth Planet Sci Lett 183:107–120

Quigley M, Sandiford M, Fifield K, Alimanovic A (2006) Bedrock erosion and relief production in the northern flinders ranges, Australia. Earth Surf Process Landf 32:929–944

Quigley M, Sandiford M, Cupper ML (2007) Distinguishing tectonic from climatic controls on range-front sedimentation. Basin Res 19:491–505

Rychagov GI (1997) Holocene oscillations of the Caspian Sea, and forecasts based on palaeogeographical reconstructions. Quat Int 41–42:167–172

Sandiford M (2003) Neotectonics of southeastern Australia: linking the quaternary faulting record with seismicity and in situ stress. In: Hillis RR, Muller D (eds) Evolution and dynamics of the Australian Plate. Geological Society of Australia Special Publication, Australia 22:101–113

Sandiford M, Quigley M (2009) TOPO-OZ: insights into the various modes of intraplate deformation in the Australian continent. Tectonophysics 474:405–416

Schmid RM (1988) Lake Torrens brine. Hydrobiology 158:267–269

Su YH, Zhu GF, Feng Q, Li ZZ, Zhang FP (2009) Environmental isotopic and hydrochemical study of groundwater in the Ejina Basin, northwest China. Environ Geol 58:601–614

Tapponnier P, Molnar P (1977) Active faulting and tectonics in China. J Geophys Res 82:2905–2930
Tuzhilkin VS, Katunin DN, Nalbandov YR (2005) Natural chemistry of Caspian Sea waters. Hdb Env Chem 55:83–108, Springer-Verlag
Wang G, Cheng G (2000) The characteristics of water resources and the changes of the hydrological process and environment in the arid zone of northwestern China. Environ Geol 39:783–790
Wei K, Gasse F (1999) Oxygen isotopes in lacustrine carbonates of West China revisited: implication for post glacial changes in summer monsoon circulation. Quat Sci Rev 18:1315–1334
Wen X, Wu Y, Su J, Zhang Y, Liu F (2005) Hydrochemical characteristics and salinity of groundwater in the Ejina Basin, Northwestern China. Environ Geol 48:665–675
Williams WD, Kokkinn MJ (1988) The biogeographical affinities of the fauna in episodically filled salt lakes: a study of Lake Eyre South, Australia. Hydrobiology 158:227–236
Yang B, Shi Y, Braeuning A, Wang J (2004) Evidence for a warm-humid climate in arid northwestern China during 40–30 ka BP. Quat Sci Rev 23:2537–2548
Zheng X (1987) Salt lakes and their origins in Xinjiang, China. Chin J Oceanol Limnol 5:172–185
Zhu GF, Li ZZ, Su YH, Ma JZ, Zhang YY (2007) Hydrogeochemical and isotope evidence of groundwater evolution and recharge in Minqin Basin, Northwest China. J Hydrol 333:239–251
Zhu GF, Su YH, Feng Q (2008) The hydrochemical characteristics and evolution of groundwater and surface water in the Heihe River Basin, northwest China. Hydrogeol J 16:167–182
Zubakov VA (2001) History and causes of variations in the Caspian Sea level: the Miopliocene, 7.1–1.95 million years ago. Water Resour 28:249–256

Chapter 11
Current Continental Base-Levels Below Sea Level, Located Close to the Sea

11.1 General

The structure, the geological history, and the hydrogeological configuration, common to most of the base levels included in this category, are the following:

(a) The base-levels are currently considerably, a few tens of meters to over 420 m, below sea level.
(b) The base levels are close to the sea, usually between several to several tens of kilometers from the nearest seashore.
(c) The base levels are part of a structural depression, a trough, a rift, or an inter-mountain basin, bounded and dissected by faults that are usually still active and responsible for basin subsidence. Some of these base levels are aligned along plate boundaries. The faults might also serve as conduits for groundwater or seawater inland flow.
(d) The base-levels, at least at their deeper portion, contain a considerable thick, hydrologically conductive sedimentary basin fill, derived from the surrounding highlands.
(e) The base-levels were connected during their geological history to the sea as part of a sea arm, which was later on disconnected from the main sea body, as a result of tectonic uplift, or sedimentary blockage. This paleo-configuration often left an imprint of a present embayment where the trough or the rift that occupied the sea arm, meets the present shoreline.
(f) Following the disconnection from the sea, the already formed continental base-levels subsided tectonically, and/or declined to their present level due to climatic, whether global or local, changes.
(g) Subsequent to the above described disconnection, a rather low groundwater divide was formed, roughly coinciding with the disconnecting feature, between the sea and the newly formed declining continental base-level. The configuration of a low divide, may enable inland subsurface seawater encroachment into the lower base-level.

U. Kafri and Y. Yechieli, *Groundwater Base Level Changes and Adjoining Hydrological Systems*,
DOI 10.1007/978-3-642-13944-4_11,

(h) Most of these base-levels are at present situated in arid or semi-arid regions, which also contributes to the low elevation of the above described groundwater divide.
(i) All these base-levels occupy hypersaline lakes, sabkhas or salt pans.

A simplified hydrological model, describing the consecutive stages that formed those endorheic base-levels, based on the entire Dead Sea Rift (DSR) Basin case (Fig. 5.2) was suggested by Kafri and Arad (1978) (see also Sects. 5.3 and 11.2.5).

In the initial stage only one base-level, in the form of a sea arm, existed. Immediately after its disconnection from the sea, a new endorheic continental base-level was formed, initially at a level identical to the marine one. At this stage, a groundwater divide was formed, schematically half way between both base-levels.

Subsequently, the endorheic base-level started to decline. As a result, the groundwater table steepened from the divide toward the ongoing declining endorheic base-level. Due to the natural tendency of the water table to regain a new moderate steady state water table slope, this process, in turn, resulted in a shift of the divide, asymmetrically, toward the upper marine base-level, followed by the lowering of the divide. It should be emphasized that such a process is feasible only in those cases where the divide's location is not structurally controlled or fixed by an elevated structure underneath. To be more specific, this could happen where the base of the confining layers, that forms the base of the aquifer, is deep enough so as to permit such a shift of the divide (Kafri and Arad 1978; Yechieli et al. 2009). Such a configuration indeed exists in troughs and rifts that contain a rather thick, hydrologically conductive, sedimentary fill.

The proposed impact of the above described hydrogeological configuration on the salination of those basins below sea level is discussed in Sect. 11.11.

The amount of available information, regarding the specific base-levels, described below, differs from one base-level to another. Therefore, the DSR base-level, which has relatively abundant information and hydrogeological data, is described in detail, serving as an analogue to other similar base-levels where less information is available.

11.2 The Dead Sea Rift Endorheic Base-Level

11.2.1 General

The Dead Sea Rift (hereafter DSR) Basin, at coordinates 31°N 35°E (Figs. 8.1 and 11.1), is a continental endorheic base-level, which occupies a surface area of some 43,000 km^2 (Bookman et al. 2004). A considerable part of the basin, within the DSR proper, lies below sea level, reaching the deepest, still declining, point on earth of ca 420 m bsl in the Dead Sea proper.

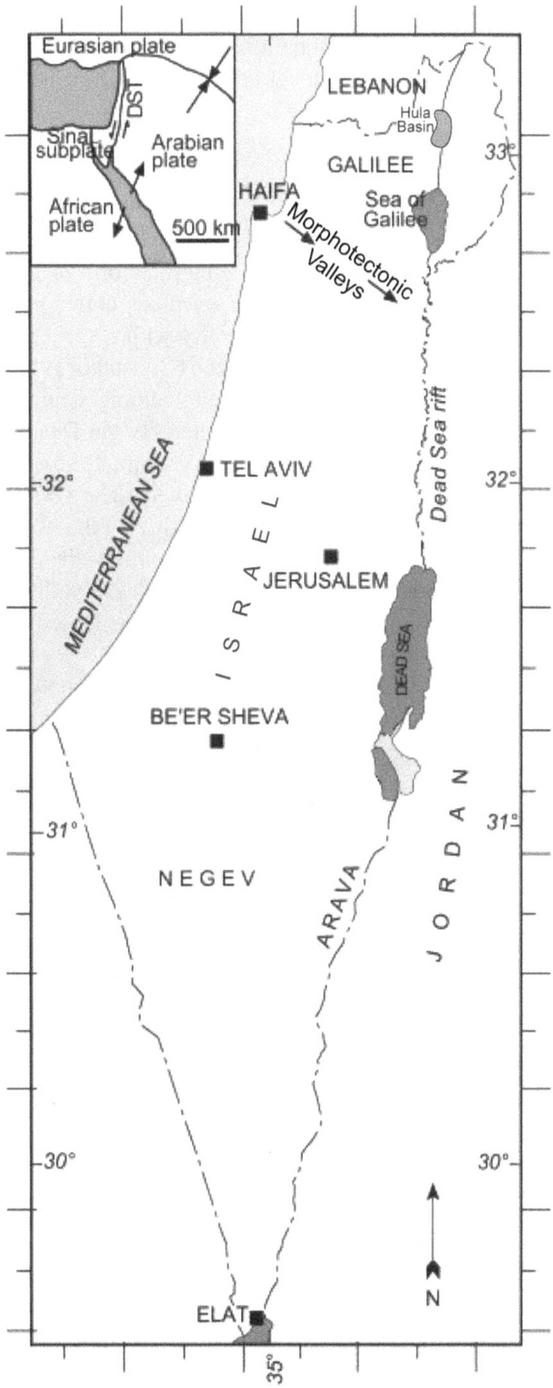

Fig. 11.1 Map of the Dead Sea Rift Basin, including the main structural features (inserted in the *upper left side*)

The geology, structure and hydrogeology of the DSR basin was studied in detail and was described in numerous studies and thus can serve as an analogue to similar base-levels below sea level in the world where less information is available.

11.2.2 Structure and Tectonics

The DSR is part of a 1,000 km long transform that links the Red Sea spreading area with the Zagros–Taurus zone of plate convergence. Along this transform the Arabian plate has moved sinistrally with respect to the Sinai sub-plate (Fig. 11.1) by some 105 km since the Neogene (Garfunkel 1981). The following rhomb-shaped grabens, which appear as morphotectonic depressions, were formed along the transform within the DSR basin, namely the Dead Sea, the Sea of Galilee and the Hula Basin (Garfunkel and Ben-Avraham 1996), presently at the elevations of ca 420 bsl, 210 bsl and 70 asl respectively. The DSR is structurally connected to the Mediterranean Sea and to the Haifa Bay by the NW–SE directed northern morphotectonic valleys of northern Israel, namely the Yizre'el and Bet Shean valleys, which are bounded by active faults. These valleys enabled the Pliocene marine ingression from the Mediterranean to the DSR (i.e., Kafri and Ecker 1964; Zak 1967) (see below).

The DSR, and the connected northern morphotectonic valleys, were subjected to neotectonic activity during the Quaternary (Garfunkel et al. 1981), a process which continues to the present, expressed by tectonic deformation and displacements, by both paleo and current seismicity and by continuous subsidence of the rift relative to its margins. Quaternary fault displacements and resultant basin subsidence, at a rate between 0.3 and 1 m/ky, were described from the Dead Sea area, by Bartov and Sagy (2004) and by Bartov et al. (2006). The roughly same rates of subsidence were noted by Zak (1967), Begin and Zilberman (1997), Horowitz (2001) and Lisker et al. (2009). Current subsidence rates, within the DSR, between 5 and 20 mm/y, were obtained by InSAR repeated measurements (Baer et al. 2002). Subsidence rates between 0.5 and 2.2 m/ky were also obtained for the Hula basin in the north (Kafri et al. 1983; Heimann and Steinitz 1989; Begin and Zilberman 1997).

Paleo, as well as historic seismicity, were discussed among others by Ben Menahem (1981), Marco et al. (1996), Ken-Tor et al. (2001) and Migowski et al. (2004). The existing current seismicity of the DSR is continuously monitored, as described by Shapira (1992) and Salamon et al. (2003).

11.2.3 The Hydrological History of the DSR Base-Level

The hydrological history of the Dead Sea Basin and its occupied paleo lakes, since the Neogene, was discussed in detail in several studies (i.e., Horowitz 2001; Stein 2001). A rather concise description of the above is given herein as follows:

During the Late Neogene, the Mediterranean Sea intruded the basin through the internal northern morpho-tectonic valleys of northern Israel, forming the Sedom Lagoon as an inland evaporitic basin (Zak 1967). The Pliocene eustatic sea level ingression, at an elevation around 50–100 m asl (Miller et al. 2005), filled the internal northern morphotectonic valleys and the Rift Valley, all the way from the present Sea of Galilee area to a distance of several tens of kilometers south of the present Dead Sea (Zak 1967; Kafri and Ecker 1964; Horowitz 2001; Stein 2001; Kafri et al. 2008). It is important to note that the Hula Basin, which is part of the Rift, was not invaded by the Pliocene transgression being separated from the Rift in the south by the elevated Korazim block. The elevation of the Hula Basin at that time was assumingly similar to its present one, namely above the Pliocene eustatic level, as is evidenced by the fact that neither marine or lagoonal saline sediments, nor trapped seawater or brines, are found within the thick basin fill (Kafri et al. 1983). Subsequently, the Sedom Lagoon was disconnected from the open sea and started to desiccate (Zak 1967; Kafri and Ecker 1964; Stein 2001; Torfstein et al. 2009). The desiccation of the Sedom Lagoon was responsible for the formation of concentrated, mostly Ca-chloride, brines which penetrated the adjoining aquifers of the DSR basin as well as to the formation of salt bodies and diapirs within the DSR (Zak 1967; Starinsky 1974; Stein 2001; Waldmann et al. 2007; Weinberger et al. 2007).

Following the disconnection of the Dead Sea Basin from the Mediterranean Sea, three known major lacustrine and hypersaline water bodies (Katz et al. 1977; Torfstein et al. 2009) consecutively occupied the basin throughout the Pleistocene–Holocene and until today. These are Lake Amora (including Lake Samra), Lake Lisan and the present remnant Dead Sea. The water levels of these lakes have risen and declined as a result of both tectonic subsidence and climate changes (Neev and Emery 1967; Stein 2001; Bartov et al. 2006; Waldmann et al. 2009).The timing of the different lakes' levels was defined by dated lacustrine as well as near shore sediments, by shore terraces (Bowman 1971) and by dated sediments that were deposited in near shore caves (Lisker et al. 2009).

Lake Amora existed at least between 740 and 70 ka whereas its inclusive younger Lake Samra existed between 135 and 70 ka. The limnological history of Lake Amora displays cyclic shifts between dry lowstands and wet highstands, which prevailed during time spans exceeding tens of thousands years. The levels of Lake Samra fluctuated between 320 and 380 m bsl (Stein 2001; Torfstein et al. 2009). The lakes' level curve of the Samra and the Lisan lakes is shown on Fig. 11.2.

Lake Lisan, discussed in numerous studies, existed between 70 and 15 ka and its levels fluctuated between 330 and 150 m bsl (Stein 2001; Torfstein et al. 2009).

The maximal extent of the Lisan Lake, within the DSR, is shown on Fig. 11.3.

The Dead Sea levels, during the Holocene, were around 400 m bsl, with a maximum fluctuation of several tens of meters (Bookman et al. 2004).

As mentioned before, elevation of cave levels yielded information with regard to their related hydrological base levels (see also Sect. 4.1.3.2). The caves in the Dead Sea area were studied mainly by Frumkin (2001) and Frumkin and Fischhendler

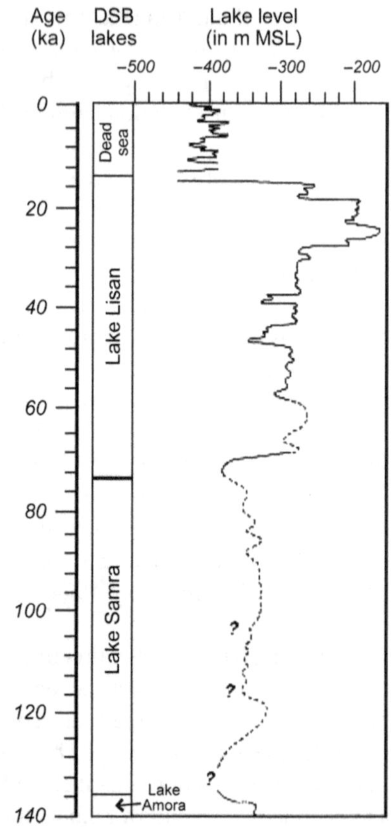

Fig. 11.2 Lake level curve of the Samra and the Lisan lakes [modified after Waldmann et al. (2009)]

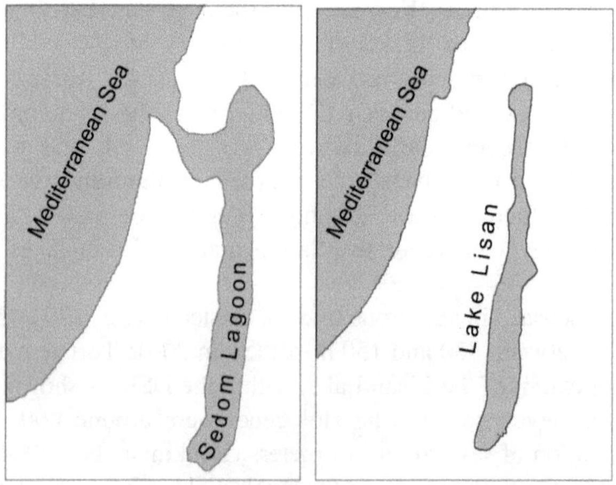

Fig. 11.3 Map showing the extend of the Sedom Lagoon and the maximal extent of Lake Lisan [after Gvirtzman (2006)]

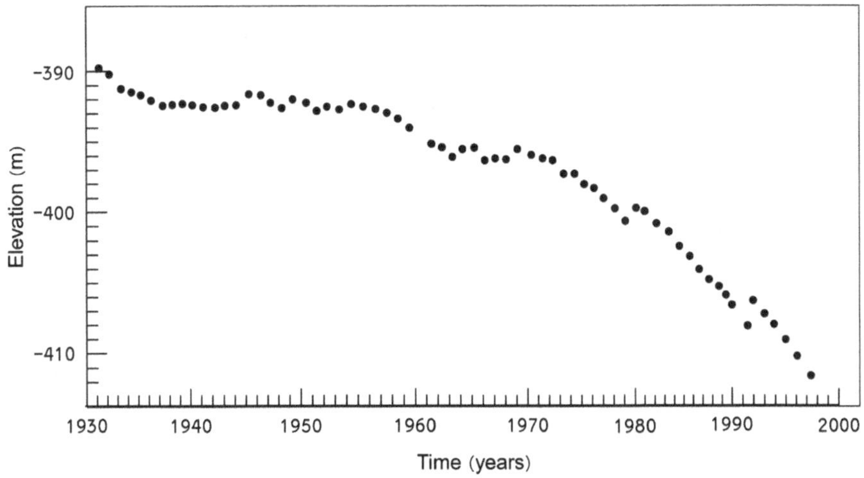

Fig. 11.4 The decline of the Dead Sea level since 1930 (data from the hydrological Service of Israel)

(2005). Lisker et al. (2009) studied the Lake Lisan levels, based on dated stromatolites found in caves on the western Dead Sea fault escarpment. They concluded that Lake Lisan levels are higher by several tens of meters than those based on lake sediments, due to the larger subsequent subsidence of the central part of the Dead Sea Basin. Kafri and Yechieli (2010) studied the role of the different paleo base-levels of the DSR basin in the formation of, and the relationship to, the sub-horizontal cave horizons aligned on the western rift margin. They also described the relationship of the different cave levels with their relevant paleo base levels.

The current Dead Sea level decline is continuously monitored in the last decades. Since the early 1960s, considerable amounts of water are diverted from the Sea of Galilee, thereby reducing the inflow of the Jordan River to the Dead Sea. In addition, the potash industries of Israel and Jordan are pumping the Dead Sea brine for production and more surface water as well as groundwater is being exploited in the basin. These resulted in a sharp drop of the Dead Sea level (Fig. 11.4) at the rate of 0.5 m/y at the beginning and around 1 m/y recently (Yechieli et al. 1998; Lensky et al. 2005). The Dead Sea, whose present (2008) level is around 420 bsl, is continuously declining and is expected to reach a new steady state at an elevation of ~550 m bsl in about 350 years (Yechieli et al. 1998).

11.2.4 Current Groundwater Flow Regime

Due to the fact that the DSR is an endorheic basin, it serves as a discharge zone to convergent surface flow, as well as to groundwater flow from an eastern groundwater divide in the mountains of Jordan and from the western divide in mountain ranges of Israel. Within the DSR, the basin is bounded to the north by a groundwater divide north of the Hula Basin, close to Mount Hermon and to the south by a groundwater divide in the Arava Valley between the Dead Sea and the Red Sea (Kafri et al. 2008). Within the DSR itself, the Hula Basin and the Sea of Galilee base-levels are intermediate, flow-through, base-levels whereas the Dead Sea proper is the terminal one. The convergent groundwater flow to the above base-levels appears as both subsurface flow into the basin fill, and as springs that ascend along the faults that bound the DSR.

The springs that issue at the margins of the Hula Basin, exhibit a normal water temperature. Those that ascend at the western margins of the Sea of Galilee and the Dead Sea basins are both of normal water temperature as well as thermal springs, indicative of their origin from relatively deep seated aquifers (Levitte and Eckstein 1979; Vengosh et al. 1991; Gvirtzman et al. 1997; Stanislavsky and Gvirtzman 1999; Shalev et al. 2007).

11.2.5 Water Salinities of the DSR Base-Levels and Their Adjoining Groundwater Systems

The hydrogeological and hydrochemical relationship between the above fresh, saline or hypersaline lakes or base-levels and their adjoining groundwater systems depends partly on their difference in salinity and density. The first Pliocene marine ingression into the DSR was assumingly, until its disconnection from the sea, typified by normal seawater salinity. Subsequently, the desiccating Sedom Lagoon was formed, depositing huge salt bodies in the Dead Sea area (Zak 1967) and more to the north, somewhat south of the Sea of Galilee (Marcus and Slager 1985). The lagoon brine was, therefore, supersaturated in regard to halite, similar to that of the present Dead Sea brine. The salinity of the Lisan Lake, the precursor of the present Dead Sea, however, was about half or less that of the present Dead Sea brine (Katz et al. 1977), exhibiting a north–south salinity gradient along the DSR (Begin 2002; Begin et al. 2004).The present Dead Sea is about ten times more saline than normal Mediterranean seawater ($\sim$220 versus $\sim$22 g/l Cl). The brine density is higher than that of the Mediterranean seawater (1.25 versus 1.027). As a result of the different water densities, the fresh–saline water interface between the fresher groundwater and the Dead Sea brine is about ten times shallower and its slope is smaller than in oceanic coastal aquifers (Yechieli 2000).

The segmentation of the DSR into three sub-basins or base-levels, namely the Hula, the Sea of Galilee and the Dead Sea proper, is also expressed by the different range of salinities of their present adjoining groundwater systems as detailed herein:

(a) *The Hula Basin* is a flow-through base-level at an elevation of approximately 70 m asl. Since the Neogene, as mentioned before, it acted as a subsiding depo-center, isolated from the southern portion of the DSR by the elevated Korazim block between the Hula Basin and the Sea of Galilee and remaining above sea level. As evidenced by the absence of any saline or evaporite sediments within the basin fill, and due to its elevation above sea level, it was concluded that it was not invaded by saline waters of neither the Sedom Lagoon or Lake Lisan (Kafri et al. 1983; Kafri and Lang 1987; Heimann and Steinitz 1989; Horowitz 2001; Kafri and Horowitz 2003; Kafri and Yechieli 2010). Groundwater within the basin fill, as well as in the adjoining aquifers is fresh, with water chlorinity around a few to several tens of milligram/liter only. This is explained by the fact that this base-level, in the past and at present, serves as a flow-through base-level.

(b) *The Sea of Galilee* or Lake Kinneret is also a flow-through base-level, downstream and assumingly down-gradient to the Hula Basin, at an elevation of approximately 210 m bsl. The lake is fed by the Jordan River and some smaller streams, as well as by convergent subsurface groundwater flow including saline springs that issue at the lake margins and from the bottom of the lake. The lake water is fresh, whereas the adjoining active groundwater systems and their related springs reveal a wide spectrum of salinities from fresh water through brackish ones and attaining a maximum value of salinity of 35 g/l, lower than that of Mediterranean seawater. Since the Sea of Galilee is one of the major water sources of Israel, the mechanism and origin of its mild salination was studied along the years by several studies. It is accepted by all the researchers that the different salinities stem from different proportions of mixtures between cyclic waters, derived from the surrounding intake areas, and a saline end member. There are two basic approaches explaining both the salination mechanism and the origin of the saline end member.

The first approach, which is mainly based on geochemical considerations, claims that the saline end member represents trapped seawater or Ca chloride brines (with salinities that far exceed those of normal seawater) that migrated in the past into the aquifers. These were subsequently diluted and flushed by cyclic waters to the DSR base-level. This approach was discussed in detail by the following:

- Goldshmidt et al. (1967) claimed that the saline sources are brines, formed by evaporation of the Pliocene marine transgression to the DSR, that are since then flushed to the base-level.
- Mazor and Mero (1969a, b) assumed that the saline end member is entrapped, slightly altered Pliocene seawater, that is presently squeezed out from the fill of the DSR as a result of tectonic compression.
- Starinsky (1974) stated that the saline waters that occupy the aquifers are mixtures of fresh water and Ca-chloride continental brines, which originated in the Dead Sea area and subsequently migrated northward along the DSR.
- Gvirtzman et al. (1997) described, using numerical modeling, the density driven flow of the above described entrapped brines into the neighboring aquifers.
- Bergelson et al. (1999) described the stages of dilution of the ancient brines and their subsequent mixture with the fresh water through flushing. They suggested that the water emerging at saline springs in the Sea of Galilee are remnants of the Lisan Lake, which was considerably less concentrated here than in its southern portion within the DSR. They differentiate between the Kinneret brine, whose

salinity did not exceed 35 g/l and is part of the active hydrological system, and the assumed deeper more concentrated brine of the Dead Sea type which is probably isolated from the currently active hydrological system.

- Hurwitz et al. (2000) also attributed the saline end member to saline Lisan Lake waters which percolated to the aquifers that bound the base-level.
- Recent studies analyzed and modeled the salination processes of the Sea of Galilee (Abbo et al. 2003; Rimmer and Gal 2003), suggesting the possibility of contribution of saline water from the deep seated sources.

A second approach was presented by Kafri and Arad (1979), who described a mechanism of current subsurface seawater encroachment from the Mediterranean Sea into the internal valleys and all the way to the DSR base-level (Fig. 11.5), on top of the mechanisms described above. It should be noted that the term "current", used herein, is differentiated from subsurface flows regimes of older paleo time spans. Still, the water, whether fresh or saline, is related to prevailing hydrological systems and base levels and may attain an age of thousands of years which stems from their flow velocities and residence times.

The proposed mechanism was based on the following hydrogeological set up that assumingly prevails in the study area (Kafri and Arad 1979; Kafri et al. 2007b): (1) Elevation difference of 200–250 m between the Mediterranean and the northern DSR base-levels and a hydrological continuity between both. (2) The existence of a wide range of salinities, not exceeding, however, the threshold value of present Mediterranean water salinity. (3) The regional JGA aquifer is in places exposed on the seashore and is hydraulically open to the Mediterranean Sea offshore in the west, and thus is subjected to subsurface seawater encroachment (Kafri 1967, 1969; Arad et al. 1975; Kafri et al. 2007a, b). (4) The base of the hydrological conductive layers, together with the fault systems between both base levels, is suggested to be low enough to allow sub-surface seawater encroachment all the way to the lower base level (5) The existence of a low, and asymmetrically located regional

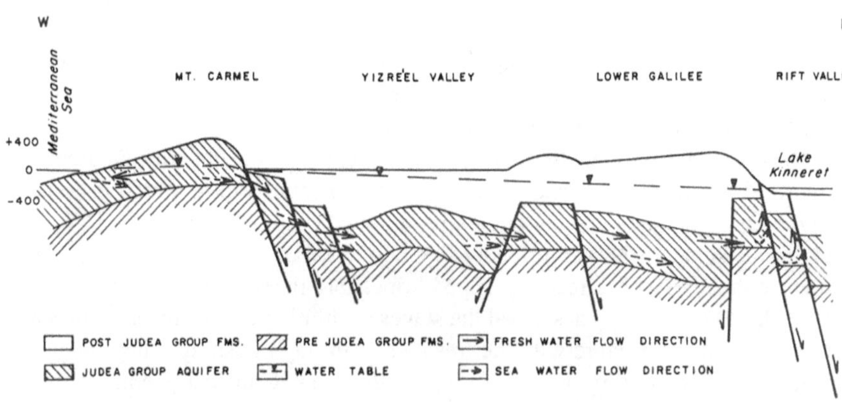

Fig. 11.5 Schematic hydrogeological cross-section from the Mediterranean Sea to the Dead Sea Rift [after Kafri (1984)]

groundwater divide between both base-levels that dictates a rather shallow interface. Such a configuration, in turn, enables seawater encroachment underneath and across the divide (Kafri and Arad 1978). Several geochemical arguments were given to support a Dead Sea type brine origin of the saline end member. Alternative interpretation, as to the origin of the saline end member was suggested by Kafri and Arad (1979). It is possible that the recent seawater encroachment serves as an additional flow mechanism that allows the ascend of assumingly older saline water.

The above proposed model assumingly enables sea water encroachment at a considerable depth into the DSR base level. The emergence of the saline end member to the surface as saline springs or as shallow sub-surface flow is possible, provided that the following conditions exist:

1. Faults that bound the rift act in places as no flow boundary and force the flow upward. Such faults exist in the western part of the DSR, acting as conduits to preferential vertical groundwater flow.
2. An already existence of a saline water body within the DSR, preferable of higher density that seawater.
3. Some of the ascending force can be provided by the head of the fresh groundwater system, recharged in the Galilee mountain area, as suggested by Goldshmidt et al. (1967). This mechanism is expected to dilute the original saline end member as is indeed found in some parts of the study area.

This approach was supported by findings of a deep electromagnetic survey that was carried out along the northern morphotectonic valleys between the Mediterranean Sea and the Sea of Galilee (Fig. 11.6) which were reported in internal reports as well as by Kafri et al. (2007a, b). The FEFLOW simulations in the latter study was setup so as to satisfy the above described conditions and thereby to enable the

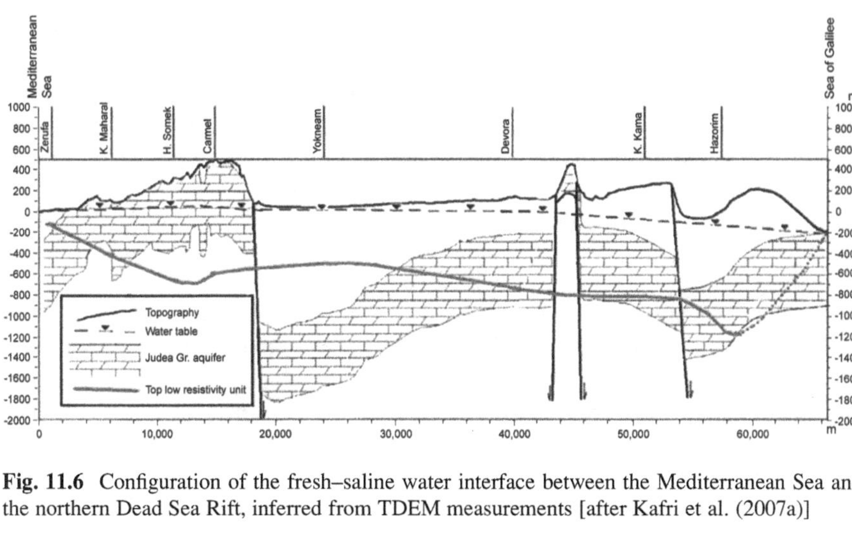

Fig. 11.6 Configuration of the fresh–saline water interface between the Mediterranean Sea and the northern Dead Sea Rift, inferred from TDEM measurements [after Kafri et al. (2007a)]

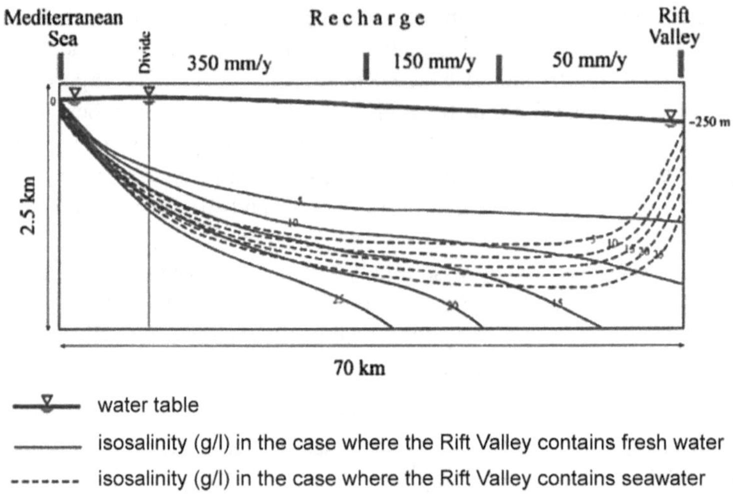

Fig. 11.7 A FeFlow simulation of groundwater flow and mass transport between the Mediterranean and the northern Dead Sea Rift base levels [after Kafri et al. (2007a)]

feasibility of the proposed mechanism (Fig. 11.7). An additional support was obtained from findings in a recently drilled borehole east of the groundwater divide between the Mediterranean and the DSR. Here, groundwater with salinities somewhat exceeding 16,000 mg/l Cl was detected at the bottom of the active Judea Group Aquifer beneath a fresher groundwater body. This water, whose geochemical affinity is quite similar to diluted seawater rather than to diluted brines, imply the process of seawater intrusion at least up to this area. The age of the saline water is not known, but it is possible that part of it is somewhat older than recent seawater due to the relatively slow flow velocity of part of the flow.

(c) *The Dead Sea* basin proper base-level serves as the terminal base-level of the system to both surface and groundwater flows. The present water salinity of the Dead Sea water is around 340 g/l. The input of surface, mainly fresh, water to the lake is from the Jordan River in the north and from streams that drain the western, eastern and southern parts of the basin. Fresh groundwater recharged in the surrounding intake areas (Arad and Michaeli 1967) as well as brackish paleo-groundwater (Issar et al. 1972; Yechieli et al. 1992) also flow in a convergent pattern to the base-level. There they mix, in different proportion, with a saline end member ending up as brackish to hypersaline subsurface flows or springs that issue at the lake's margins. The origin and geochemistry of the saline end member was analyzed and discussed in several studies (i.e., Starinsky 1974; Yechieli et al. 1996; Klein-BenDavid et al. 2004; Gvirtzman 2006; Waldmann et al. 2007; Katz and Starinsky 2009). According to the above, the saline end member that was originated in the Pliocene Sedom Lagoon and the consecutive lakes, are Ca-chloride brines, formed in a marine lagoon or in an inland more concentrated saline lake. These brines were subjected, during the history of their evolution, to periods of inter-mixing, processes of water rock interaction, evaporation, and deposition of

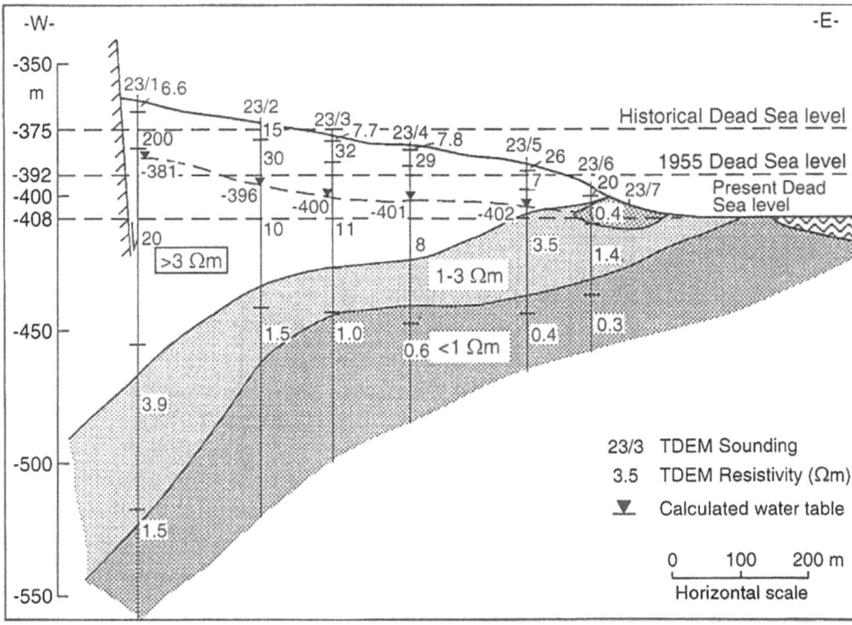

Fig. 11.8 A TDEM cross-section in the coastal aquifer of the Dead Sea exhibiting the fresh–saline water interface configuration and the overlying already flushed zone [after Kafri et al. (1997)]

evaporates, such as halite, as well as of "fresher" water sediments, such as aragonite. The brines are penetrating the adjoining groundwater systems by a density-driven mechanism and are mixed with the fresher groundwater. The current interface, or interfaces in the case of a multi aquifer system, between the brines and the overlying fresher groundwater is shallower and of a milder slope, as compared to a case of normal seawater according to the water density differences (Yechieli 2000). The interface is being detected by borehole water sampling, borehole electrical logging and surface time domain electromagnetic (TDEM) measurements (Figs. 11.8 and 11.9) (Kafri et al. 1997; Yechieli et al. 2001, 2010; Yechieli 2006; Batayneh 2006). The configuration of current and paleo interfaces, related to the discussed paleo lakes, were also suggested, based on the TDEM method, from the northern Arava Valley south of the Dead Sea (Fig. 11.10) (Kafri et al. 2008).

11.2.6 The Response of the Groundwater System to the DSR Base-Level Changes

11.2.6.1 General

The different aspects of the current groundwater system response to the DSR base-level changes are detected and monitored. Those related to paleo groundwater

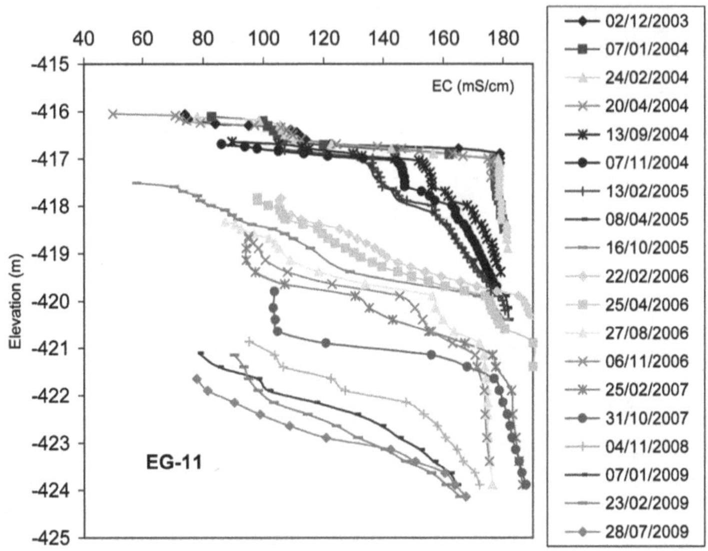

Fig. 11.9 The location of the fresh–saline water interface as detected by electrical conductivity (EC) profiles in observation borehole in the Dead Sea coastal aquifer [after Yechieli et al. (2010)]

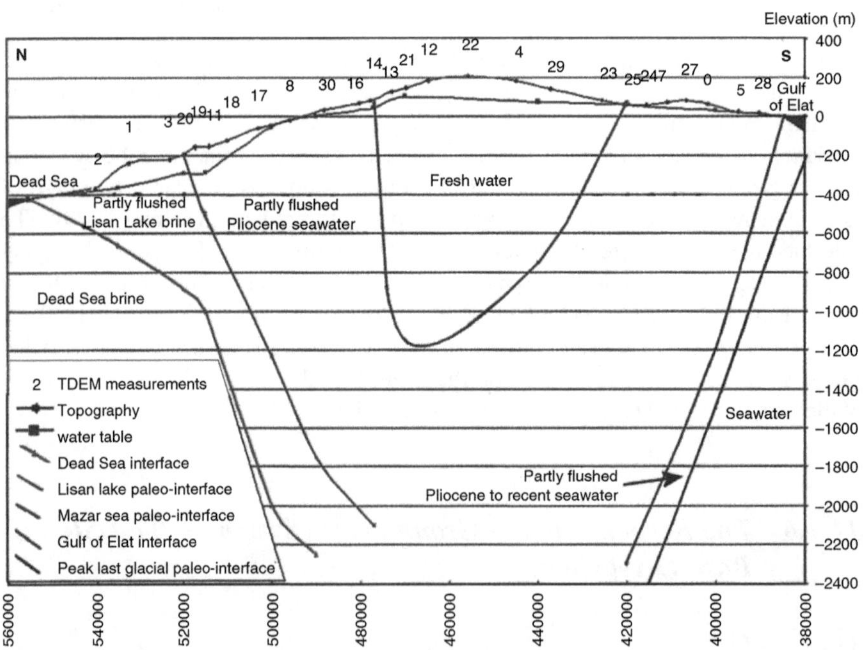

Fig. 11.10 Current and paleo fresh–saline groundwater interfaces inferred from TDEM measurements between the Gulf of Elat (Red Sea) and the Dead Sea base levels [after Kafri et al. (2008)]. The number below the horizontal axis are the Israeli coordinates

systems could be reconstructed by supporting evidence or by simulation. Future behavior of the system was also simulated. This includes the following aspects: future predicted base-levels changes, changes of groundwater levels, changes in the location of groundwater divides, changes in groundwater flow to the base-levels, changes of groundwater salinity and interface location and flushing processes.

11.2.6.2 Paleo-Groundwater System

The past response of the groundwater system is deduced, based on existing direct or supporting evidences, on logical assumptions, and on simulations.

Regarding the Hula Basin in the north, it is assumed that the elevation of that base-level remained, since the Neogene, roughly at the same elevation above sea level due to the same rate of subsidence and sedimentation (Kafri et al. 1983). It is, thus, assumed that no drastic changes in the groundwater regime and/or salinity occurred during that time span.

Regarding the central and southern portions of the DSR, namely the Sea of Galilee and the Dead Sea proper basins, it is assumed that prior to the formation of the DSR as a base-level, groundwater flow occurred from east of the present day DSR all the way to the paleo Mediterranean base-level, depending on the specific geological structures in between. Subsequent to the formation of the DSR as a base-level, and its intrusion by the Pliocene marine ingression through the northern morphotectonic valleys, it is assumed that the groundwater flow pattern adjusted itself to the newly formed base-level configuration. The disconnection of that marine arm from the sea, formed the inland hyper-saline and continuously declining Sedom Lagoon base-level and thereby created, supposedly, a new groundwater flow pattern as described and schematically shown (Fig. 11.5) by Kafri and Arad (1979). An eastward steep groundwater level was formed, that followed the base-level drop. The consequent adjustment of the groundwater flow gradient to a more gentle one, in order to gain a new equilibrium, lowered and shifted the groundwater divide asymmetrically to a location, closer to the Mediterranean base-level. That could have happened only where the underlying structure permitted it, namely along the northern morphotectonic valleys. This process, in turn, captured an additional intake area to the east (Kafri and Arad 1978). It is assumed that the different eastward paleo-groundwater levels followed their related base-levels (Amora and Lisan lakes) as evidenced by a series of caves levels aligned along the DSR western margin, that are interpreted as indicating the paleo groundwater discharge level to the old base-levels (Kafri and Yechieli 2010).

Simulation of the hydrological regime during the Lisan Lake base-level time, when the lake level reached a highest elevation around 180 m bsl, using a FEFLOW modeling software, managed to reconstruct the paleo-groundwater level, the paleo-groundwater divide location, the paleo flow to the base-level and the interface location and its configuration, tied to the relevant climate conditions and Lake Lisan salinity (Yechieli et al. 2009). It should be mentioned that, different from the northern morphotectonic valleys, this simulation was carried out for that portion of the DSR where the structure that underlies the mountain crest to the west basically dictates the divide's location. It was found that the divide shift, due to the above, was only some 400 m east of the present divide being limited by the underlying structure and thus groundwater flow to the eastern base-level was somewhat less. The fresh–saline water interface was found to be some 2–3 kms west of the present one, in accordance to the location of the paleo shorelines (Yechieli et al. 2009). The slope of the interface in this case was steeper, as compared to the present day situation, due to a smaller density difference.

11.2.6.3 Current Groundwater System

The Hula Basin does not exhibit at present any considerable base-level changes and is not subjected to any considerable salination processes and, thus, the groundwater system is approximately at a steady state.

Base (lake) level changes of the Sea of Galilee are effected by anthropogenic diversion of water from the lake and are at present limited to a few meters only between the threshold levels of 208 and 214 m bsl. As mentioned before, it is agreed that the brackish to saline groundwater represents a mixtures of a fresh and a saline end member. The hydrological mechanism, which is responsible to groundwater salinity changes, is debatable. One concept claimed that the saline end member, beneath the Sea of Galilee, has its own potential and thus a lowering of the lake level results in increased salinity (Mazor and Mero 1969a, b; Simon and Mero 1992). An alternative concept by Goldshmidt et al. (1967) claimed that the saline end member is driven in the margin of the basin upward by the head of the adjoining groundwater system and thus lower groundwater levels, whether natural or caused by exploitation, are supposed to reduce groundwater and lake salinity and vice versa. In fact, it was found that changes in groundwater levels affect differently the salinity of different springs that issue at the margins of the lake. It is noticed that in rainy seasons, accompanied by rising lake and groundwater levels and higher spring discharges, solute concentrations decreased in the Tabgha springs but adversely increased in the Fuliye springs, suggesting a different salination mechanism to the two groups of springs (i.e., Gvirtzman et al. 1997).

In the Dead Sea Basin, the effect of the current continuous decline of the base-level, since the 1960s, on the different aspects of the groundwater regime, was monitored and described in the last decades. The effect of the drop of the lake (base) level on the adjoining groundwater levels was analyzed and discussed by Kafri (1982), Yechieli (1993, 2006), Yechieli et al. (1995, 2001, 2009) and Kiro et al. (2008). According to the above, it is recognized that the continuous drop of the lake level, interrupted by short periods of level rise, was accompanied by parallel groundwater level changes (Fig. 11.11). Moreover, it was found that the dimensions of the response are affected by both aquifer properties and the distance of the monitored site from the receding shoreline as follows (Yechieli et al. 1995, 2010; Kiro et al. 2008). The drop of the groundwater level was similar to that of the lake level in sites close to the shoreline and smaller in more distant sites. In addition, a

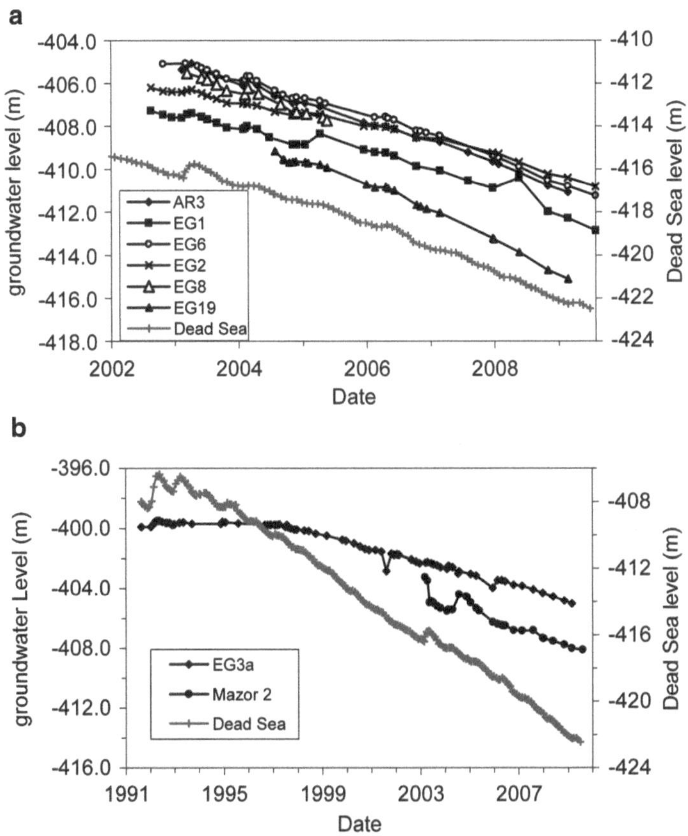

Fig. 11.11 The drop of the groundwater table in several boreholes in the Dead Sea coastal aquifer, compared to the decline of the Dead Sea level [after Yechieli et al. (2010)]. (**a**) Groundwater decline similar to that of the lake level (Wadi Arugot); (**b**) groundwater decline is smaller than that of the lake, due to greater distance from the shore and lower hydraulic conductivity of the aquifer

more considerable inland response of groundwater level drop was felt in zones of a higher aquifer permeability (Fig. 11.11).

Salinity changes in the coastal aquifer of the Dead Sea Basin, following the continuous drop of the lake level since the 1960s, were revealed by changes in the location and depth of the fresh–saline water interface(s), as a result of the groundwater level drop, as schematically shown in Fig. 11.12. Those were detected by TDEM resistivity measurements, by borehole sampling and by borehole electric conductivity (EC) repeated profiling (Figs. 11.9 and 11.13), (Kiro et al. 2008; Yechieli et al. 2010). The detection by TDEM of the current (1997) interface and the previous higher one, related to the higher (1960s) lake level by the un-flushed saline remnants in the unsaturated (vadoze) zone (Fig. 11.8), were described by Kafri et al (1997), Yechieli et al. (2001) and Yechieli (2006). In the course of this decline, the portion of the aquifer occupied previously by saline water was replaced

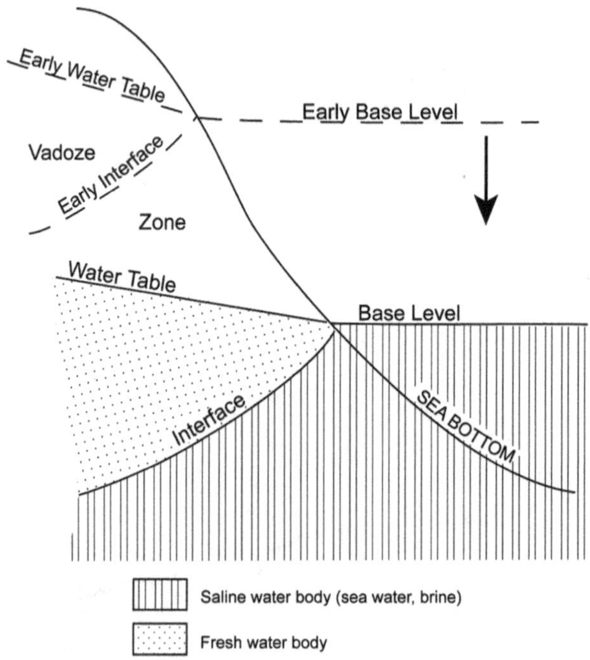

Fig. 11.12 A scheme showing the resultant drop of groundwater table and fresh–saline water interface due to the decline of an adjoining saline base level

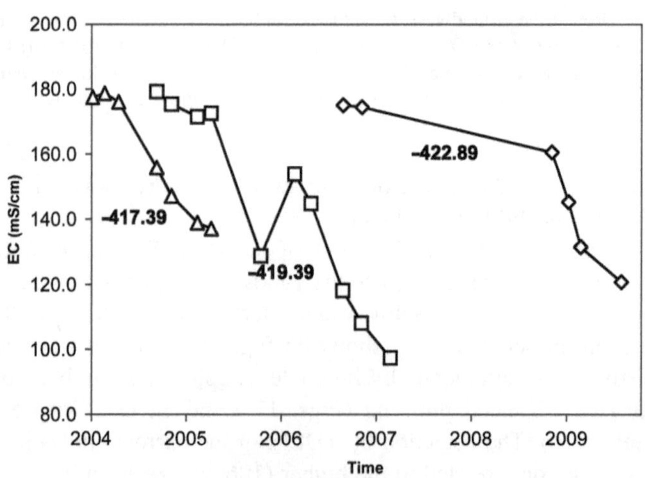

Fig. 11.13 The change with time of EC values at specific depths of the monitoring borehole, exhibiting the drop of the fresh–saline groundwater interface in the Dead Sea coastal aquifer and the effect of partial flushing of the original brines [after Yechieli et al. (2010)]

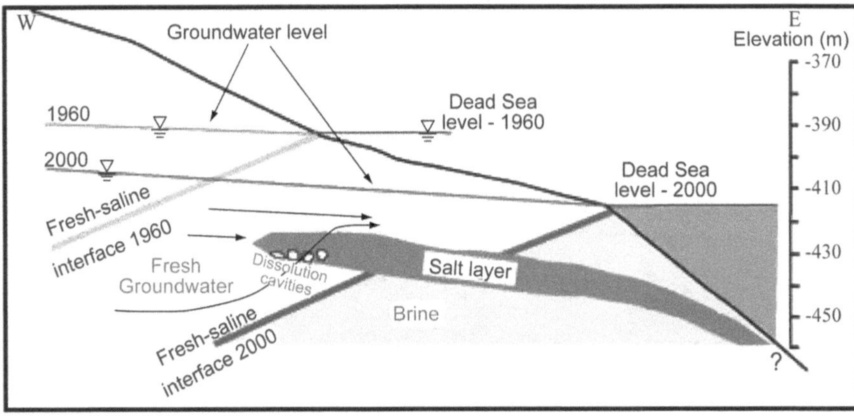

Fig. 11.14 A conceptual model showing the decline of the Dead Sea and the resultant sinkholes formation [after Yechieli et al. (2006)]

within a relative short time and flushed by fresher water. The drop of both lake level and the adjoining groundwater level resulted in a dewatering process of the upper portion of the aquifer, which was diverted into a vadoze zone. The rate of almost complete flushing of the dewatered zone was found by the TDEM surveys to be rapid, within a few decades.

An additional effect of the lake level changes could be on the discharge of the thermal springs that exist at the lake's margins. Hydrological simulations have shown (Shalev and Yechieli 2007) the disturbed equilibrium between the lake level and the springs' discharge, following lake level changes. A lake level decline results in increased spring discharges whereas a rising lake level reduces the discharge or results in the complete chocking and disappearance of the springs.

An additional by-product of the current decline of the Dead Sea and the groundwater levels is the continuing formation of sinkholes. The hydrogeological setup and the mechanism responsible for the sinkhole formation was summarized by Yechieli et al. (2006) (Fig. 11.14) as follows: The sinkhole formation occurs due to the dissolution of a 10,000 year old salt (halite) layer buried at a depth of a few tens of meters. This layer was below the brine–fresher water interface, prior to the groundwater level and associated interface decline. The decline of the interface "exposed" this salt layer to dissolution by fresher water, that is below the saturation threshold of halite, and sinkholes started to form.

11.2.6.4 Future Groundwater System

The future behavior and response of the groundwater system to the DSR base-level changes are obtained by simulations based on reasonable forecast of future base-level elevation. Since no base-level changes are expected in the foreseen future regarding the Hula Basin and the Sea of Galilee, the groundwater systems there are expected to behave roughly similar to the current situation.

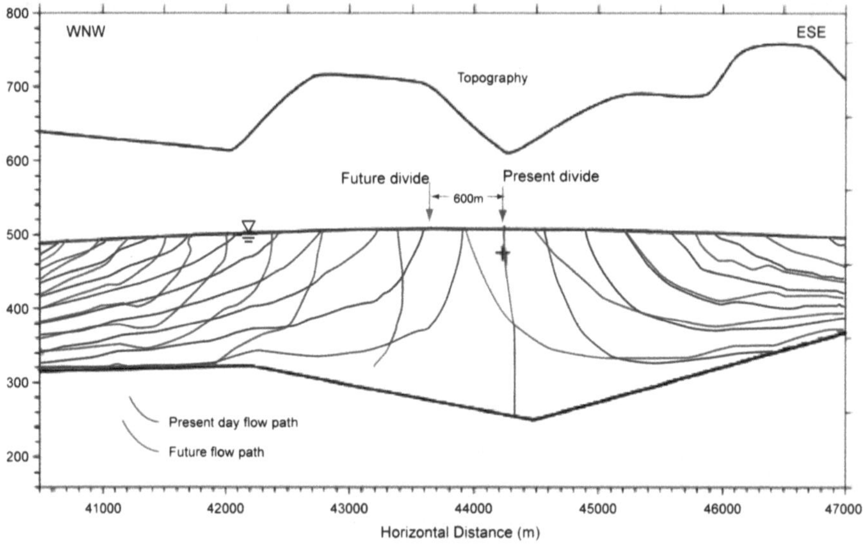

Fig. 11.15 Simulated divide shift between the Mediterranean and the Dead Sea Rift base levels as a result of future Dead Sea level decline [after Yechieli et al. (2009a)]

However, the situation is different concerning the Dead Sea Basin which is expected to continuously decline at a decreasing rate until gaining an equilibrium between input and evaporation at an elevation of approximately 550 m bsl, approximately within the next 300 years (Yechieli et al. 1998). A FEFLOW simulation of the above future scenario for the next several hundred years, yielded the following results as to the response of the groundwater system (Yechieli et al. 2009):

(a) The groundwater level will follow the lake's level drop at a decreasing drop with distance from the lake (Fig. 11.11).
(b) A westward shift of the divide by only several hundred meters is expected, since such a shift is limited by the structural axis of the mountain crest in the west (Fig. 11.15).
(c) Discharge to the base-level will considerably increase in the beginning, on the expense of storage dewatering, but will become moderate, later on, until the discharge is stabilized at values higher by some 3% than the initial ones due to the only small groundwater capture by the divide shift (Fig. 11.16).
(d) The fresh–saline interface is expected to shift some 2 km eastward of its present location (Fig. 11.17).

It is interesting to note that circulation of the intruding Dead Sea water into the aquifer is expected to continue even at this rapid drop of the present hydrological system, as exhibited by SUTRA simulations (Kiro 2006; Kiro et al. 2008) (Fig. 11.18). The drop of the Dead Sea level reduces the intruding flow into the aquifer until a new quasi steady-state is reached and the flow of both fresh and saline groundwater return to be similar to their initial previous situation.

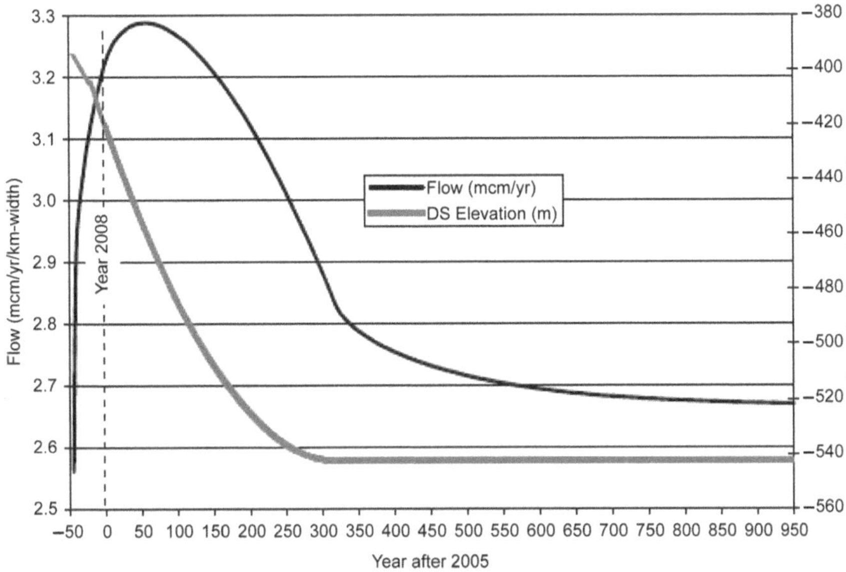

Fig. 11.16 Simulated groundwater flow to the Dead Sea due to the future decline of the sea level [after Yechieli et al. (2009)]

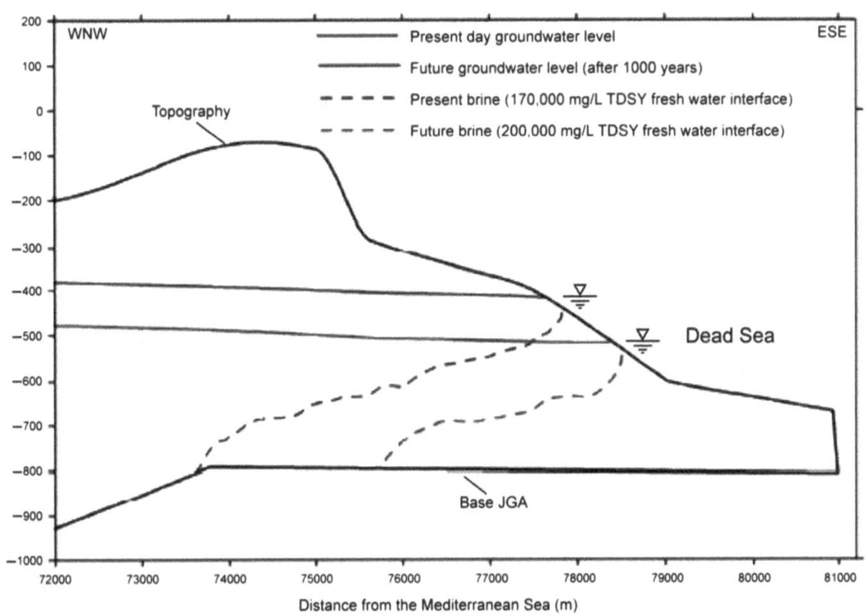

Fig. 11.17 Simulation of the present fresh/saline groundwater interface and the future (after 1,000 years) one following the predicted decline of the Dead Sea [after Yechieli et al. (2009)]

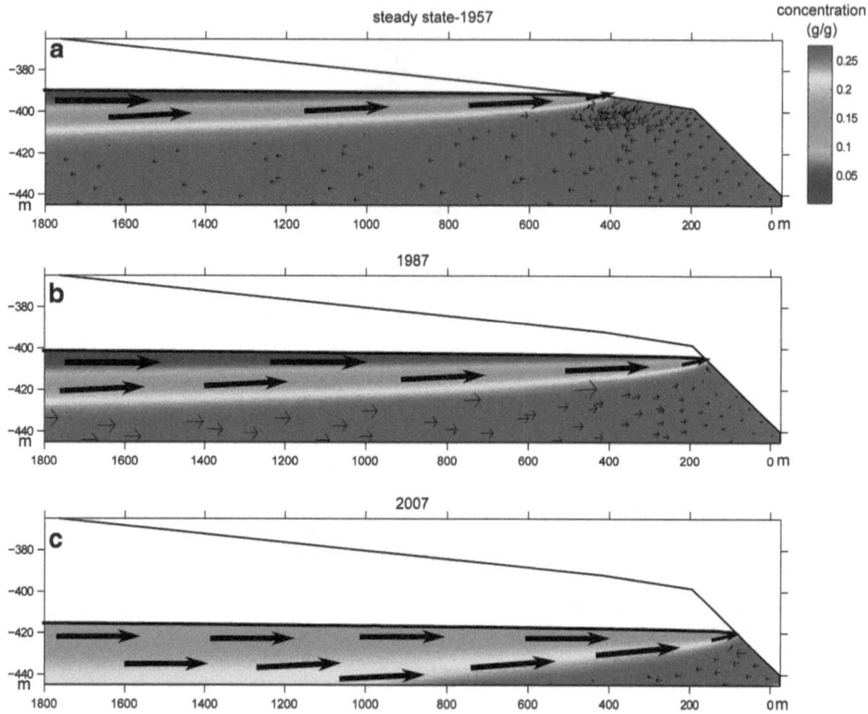

Fig. 11.18 The effect of the drop in the Dead Sea level on the fresh–saline water interface in adjacent phreatic aquifer (Kiro 2006). The *arrows* denote groundwater velocity. However, the arrows at the interface and above are not to scale. Note that at the steady state situation there is circulation of the Dead Sea water in the aquifer, while after 30 years (1987) the flow is toward the Dead Sea in the whole section. The flow to the Dead Sea below the interface resume again after 20 years

11.3 The Afar Depression-Lake Asal

The Afar, or Danakil Depression (Fig. 8.1), is a huge endorheic closed basin, considerably below sea level in the Horn of Africa, belonging to Ethiopia, Eritrea and Djibouti. The depression contains a series of lakes and salt pans at various depths below sea level: Lake Badda (−50 m), Lake Assale (−118 m), Lake Bakili (−120 m), Lake Afrera (−111 m), Lake Acori (−94 m) and the lowest Lake Asal (−155 m). The area is extremely arid with average rainfall below 100 mm/y and evaporation of about 5 m/y (Faure et al. 2002).

The depression is located within the Afar triangle which is a triple junction between the Red Sea, the Gulf of Aden and the East African, or Main Ethiopian Rift systems (Fig. 11.19). The region is a plate boundary, part of the Afro-Arabian system, which is subjected to sea floor spreading, rifting and recent tectonics, volcanism and hydrothermal activity.

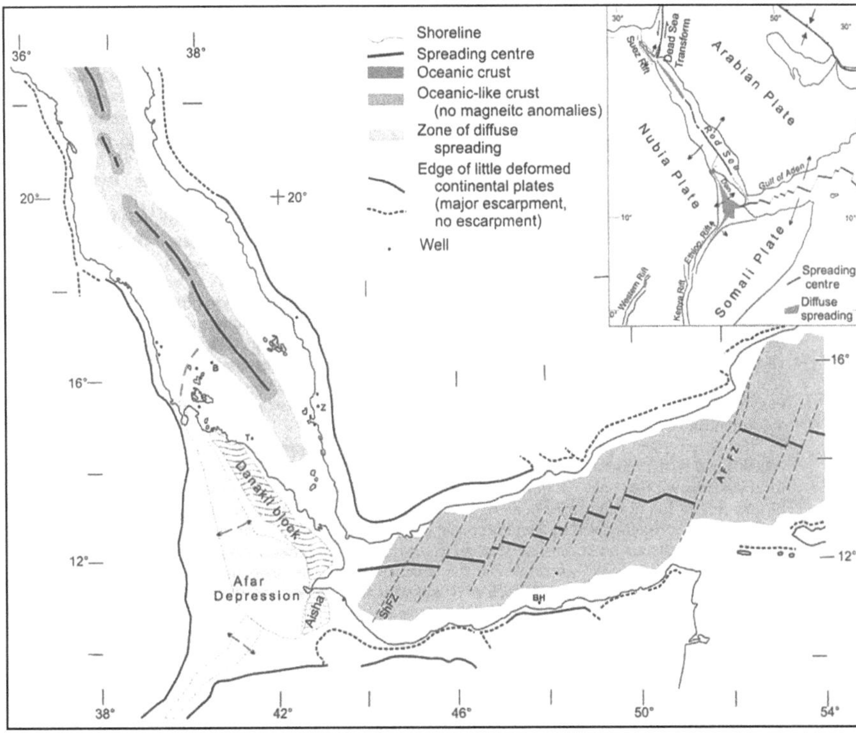

Fig. 11.19 Map of tectonic setting of the Afar Depression [after Garfunkel and Beyth (2006)]. Permission to use this figure was granted by the Geological Society of London

The structural setup and the young and current tectonics of the region were discussed in numerous studies. The triple junction and the rift systems of the region (Fig. 11.19) were described among others by Barberi et al. (1975), Garfunkel and Beyth (2006) and Buck (2006). The sea floor spreading mechanism of the Red Sea system and the plate separation were described by Allard et al. (1979) and Ayele et al. (2006). The process is being accompanied by young and current tectonics, by considerable slip rates between a few and several tens of millimeter/year, by current seismicity, by volcanism and rift segmentation, which is accompanied by fissure openings. All these phenomena were described in detail by Allard et al.(1979), Tarantola et al. (1980), Audin et al. (2001), Manighetti et al. (2001), Wright et al. (2006), Pizzi et al. (2006), Ayele et al. (2006), Buck (2006), Garfunkel and Beyth (2006) and Doubre and Peltzer (2007).

The landward extension of the eastern segment of the transform of the triple junction, namely the Gulf of Aden, forms the Gulf of Tadjourah or Ghubbet el Kharab embayment. Similarly, the landward extension of the Red Sea spreading segment into the northern Afar is manifested by the Zula Gulf on the Red Sea.

The hydrogeology of this regional depression and its hypersaline lakes, which serve as terminal base-levels, and especially the lowest Lake Asal, attracted the interest of several authors. The hydrology of Lake Asal and its adjoining groundwater system may represent and typify the entire Afar continental base-level.

Lake Asal, at coordinates 11°N 42°E, presently at 155 m bsl, is a hypersaline lake with an adjoining salt plain. The total salinity of its waters is around 300 g/l, 8–9 times more concentrated than seawater, and close to saturation with respect to halite. Mean annual rainfall in the area is 175–200 mm/y and evaporation exceeds 3.5 m/y (Gasse and Fontes 1989). The lake is fed by groundwater (Gasse et al. 1980) and is separated from the gulf of Aden and the Ghubbet el Kharab marine base-level by a 11–12 km long, and rather shallow, topographic divide, which is cut across by the tectonically active Afar Rift. As a result, the entire region and specifically the divide, is subjected to fissure openings and enlargement of fracture permeability (Allard et al. 1979; Kafri 1984; Gasse and Fontes 1989; Mlynarski and Zlotnicki 2001; Doubre and Peltzer 2007).

The paleohydrology of Lake Asal was described in detail by Gasse and Fontes (1989). According to them (Fig. 11.20), Lake Asal was almost dry in the late Pleistocene, due to the arid conditions that prevailed at that time. A highstand lake level at 160 m asl existed in the early Holocene, between 8.6 and 6 ka, and the lake at that time drained to the Ghubbet el Kharab, across a shallow sill at an elevation of some 4 m bsl, to the Gulf of Aden, both as surface and subsurface flows. Subsequent to this time span, the lake level dropped considerably to 150 m bsl, enabling at the first time subsurface seawater infiltration from the sea to the lake across the fractured divide. Since then, the lake water salinity is controlled, mainly, by seawater input and high evaporation, responsible also for Ca-sulfate and halite deposition. Similarly, paleo-lake levels between 96 and

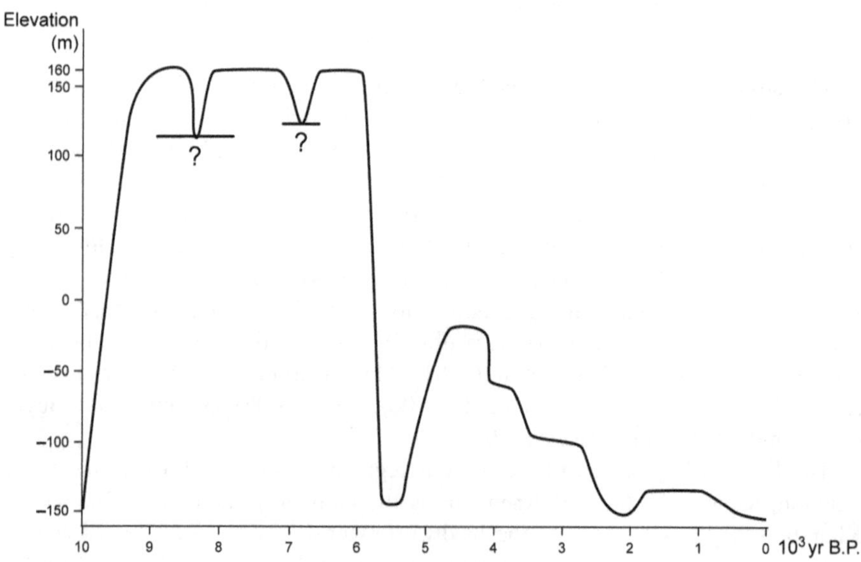

Fig. 11.20 Lake Asal level fluctuations since 10,000 years B.P [modified after Gasse and Fontes (1989)]

81 m bsl, related to the 10–7 ka time span, were described by Gonfiantini et al (1973), from Lake Afrera (at coordinates 13°N 40°E) more to the north, which is presently at 111 bsl. Lake Afrera, similar to Lake Asal is also a groundwater fed lake (Gasse et al. 1980).

The characteristics of both, the lakes and their adjoining groundwater systems, in the discussed region, were analyzed, based on their chemistry as well as on their stable isotope composition. The lakes and the groundwater systems exhibit a mixture of fresh meteoric waters and a saline end member of marine origin, described by the following: Fouillac et al. (1989) described the water composition as thermal waters close to marine composition and Sanjuan et al. (1990) described them as a mixture of thermal waters and seawater. Gasse et al. (1980) and Gasse and Fontes (1989) claimed that the main input to the system is seawater, supported also by the occurrence of the thermal Manda spring waters, of marine origin, that issue on the shores of Lake Asal. Faure et al. (2002) related the hot springs waters of the Dallol region to heated fresh groundwater that dissolved evaporites, accompanied by considerable evaporation, whereby gaining their hypersaline concentration.

The lakes water and the groundwater composition, thus, require a subsurface seawater intrusion into the lower Lake Asal base-level across the divide, that separates it from Ghubbet el Kharab Gulf. Such a configuration and mechanism, utilizing the base-levels differences and the fractured divide was suggested for Lake Asal and the Afar Depression base-level by Kafri (1984). Supporting evidences to subsurface seawater intrusion were obtained later by the following:

(a) Gasse and Fontes (1989) described, as mentioned before, the Manda spring of marine origin, on the shores of Lake Asal.

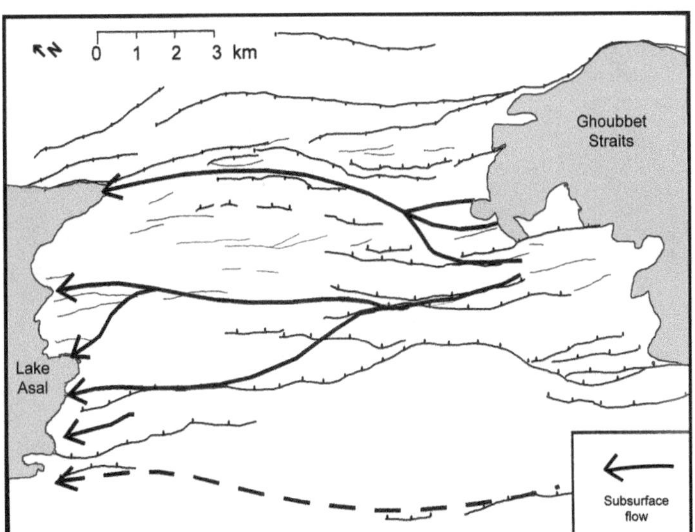

Fig. 11.21 Subsurface fluid circulation pattern from the sea inlet to Lake Asal based on SP and Telluric–Telluric prospecting [modified after Mlynarski and Zlotnicki (2001)]

(b) Houssein and Jalludin (1996) described the existence of a fresh/seawater interface within the Djibouti aquifer, some 4 kms inland at a depth of 35 m bsl.
(c) Mlynarski and Zlotnicki (2001) described, using SP and Telluric–Telluric prospecting, two main positive anomalies, parallel to the rift which are separated by a relatively large negative anomaly, exhibiting the subsurface fluid circulation pattern between the sea and the Asal Rift (Fig. 11.21). They were interpreted as exhibiting large and rapid subsurface seawater fluid transfer from the sea to the lake along permeable fault zones or beds, being warmed during their advance by the underlying geothermal flux.

It is suggested herein, as previously done (Kafri 1984), that the same similar setup might prevail in other locations in the northern Afar, enabling subsurface seawater intrusion into the Afar Depression base-level from the Red Sea.

11.4 The Qattara Depression

The Qattara Depression (Fig. 8.1), in the western desert of Egypt, at coordinates 30°N 28°E (Fig. 11.22), is a continental endorheic base-level several tens of kilometers south of the Mediterranean Sea. The surface area of the depression is close to 20,000 km^2, and it reaches an elevation of 134 m bsl, being the largest and deepest terminal base-level in the Sahara Desert. The climate is extremely arid, typical of high temperatures and annual rainfall of 25–50 mm.

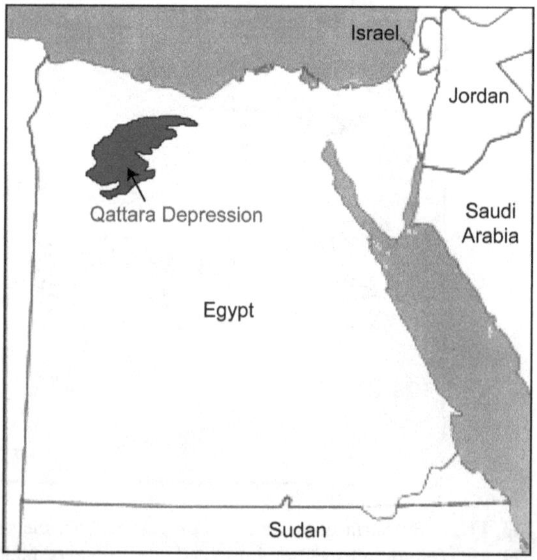

Fig. 11.22 Location map of the Qattara Depression

The geology of the discussed region, and especially the area between the Mediterranean Sea and the depression, was described among others by Said (1962), Thiele et al. (1970), El Shazly et al. (1976), Kafri (1984) and Aref et al. (2002). The general geological setting between the Qattara Depression and the Mediterranean Sea is exhibited in a north–south directed hydrogeological cross-section (Fig. 11.23). The area between the sea and the depression consists of the El Diffa Plateau, that dips to the north toward the Mediterranean Sea, and forming a topographic divide, in between, at an elevation of 230 m asl. The El Diffa Plateau basically consists of the lower Miocene detrital Moghra Formation, overlain by the middle Miocene calcareous Marmarica Formation. The bottom of the Qattara Depression and its margins are covered by sabkhas, salt pans and evaporite bearing sands and gravel.

The mechanism of the formation of the depression, to a considerable depth below sea level, is still debatable. There are yet, apparently, no structural indications as to any relationship to tectonic subsidence. Said (1962) suggested a mechanism of an erosive incision in the jointed carbonates underneath the depression to the depth of the water table. Albritton et al. (1990) proposed an initial phase, whereby the area was drained during the late Miocene through a paleo channel to the receding Mediterranean Sea. A further deepening of the area was related to mass wasting, karstic and fluvial processes as well as to wind deflation. Aref et al. (2002) adopted the idea of the above described processes, that started in late Miocene, and emphasized the role of salt weathering by mechanical disintegration accompanied by the removal of the debris by wind deflation.

The hydrology of the depression is basically dominated by its being a base-level and a terminal discharge zone to convergent groundwater flow from the south and the north. The southern flow component is the flow in the widely discussed Nubian Sandstone Aquifer, ranging in age from Cambrian to Late Cretaceous and occupying most of North Africa. Similar to other continental base-levels and oases in North Africa, the depression serves as a discharge zone to the Nubian Sandstone

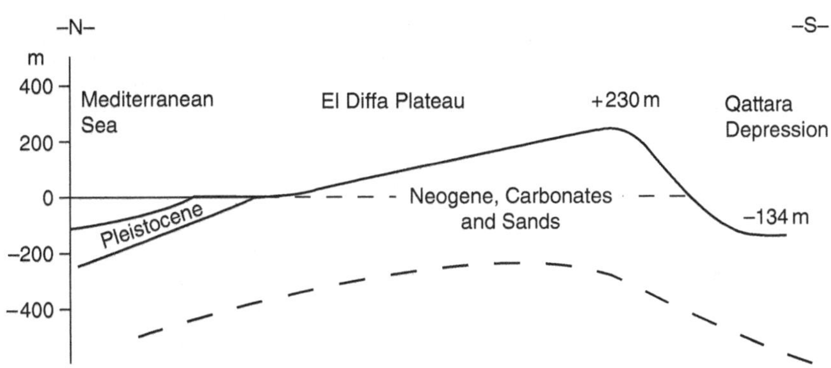

Fig. 11.23 N–S directed schematic hydrogeological cross section between the Qattara Depression and the Mediterranean Sea at a distance of several tens of kilometers [after Kafri (1984)]

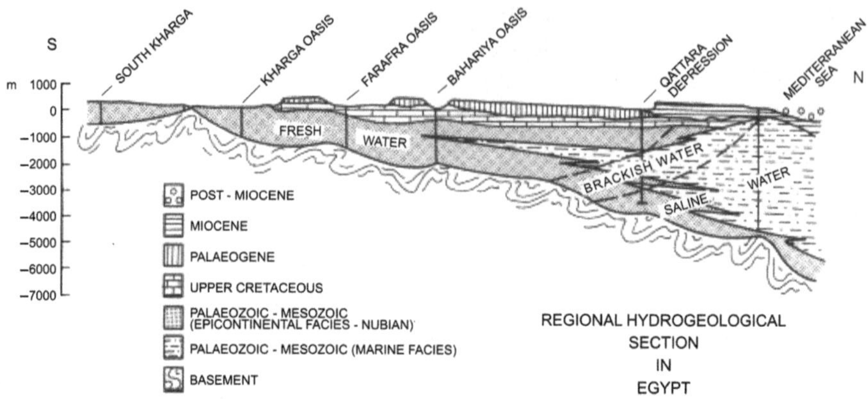

Fig. 11.24 A regional hydrogeological cross-section between the Mediterranean Sea and the Kharga springs through the Qattara Depression, exhibiting a fresh–saline water interface extending to the Qattara Depression [after Shata (1982)]. Permission to use this figure was granted by the Geological Society of London

Aquifer as noted, among others, by Shata (1982), Wright et al. (1982), Heinl and Brinkmann (1989), Lloyd (1990) and Gossel et al. (2004). Most of the Nubian Sandstone waters are paleo-waters that reveal apparent ages, based on radiogenic isotopes, of up to one million years (i.e., Patterson et al. 2005; Froehlich et al. 2007). Groundwater salinity increases from fresh waters, near the intake areas in the south, to more brackish and saline ones, closer to the discharge zone in the north. Shata (1982) suggested the existence of a fresh–saline water interface only within the Nubian Sandstone aquifer, between the Mediterranean Sea and the Qattara Depression, which does not extend farther inland into the overlying shallower aquifers (Fig. 11.24).

The north–south flow component to the depression, through the Moghra and the Marmarica formations, can be deduced from the geological setup (Fig. 11.23), as well as from water table data. Thiele et al. (1970) described an inclining water table to the south evidenced by a water table, at 6.5 m bsl and 16.6 m bsl at respective distances of 15.5 and 22 kms from the Mediterranean coastline and a water table level, more to the south, at 32 m bsl, in the northern margin of the Qattara Depression. Similarly, El Shazly et al. (1976) have shown a southward inclining water table from 7 m bsl in the north to 50 m bsl in the south, close to the depression. This water table configuration, assumingly, requires the existence of a low groundwater divide between the Mediterranean Sea and the Qattara Depression, as was suggested by Kafri (1984).

Groundwater salinities are extremely high at the bottom of the depression and beneath the salt pans, as a result of the aridity and high evaporation. El Shazly et al. (1976) and Aref et al. (2002) described salinities of up to 300 g/l.

The suggested salination mechanism of the depression seems to be similar to that of the other base levels below sea level and close to the sea, as discussed in detail in Chap. 6.

11.5 The Salton Trough Base-Level

The Salton Trough and its occupied Salton Sea, at coordinates 33°N 115°W (Figs. 8.1 and 11.25), which is presently at around 70 m bsl, serve as a continental terminal base-level. The Salton Sea is located some 150 kms north of the Gulf of California. The Salton Sea was created in 1905 following heavy winter storms and flooding of the area by the overflows from the Colorado River irrigation canals system. The area is arid with temperatures between 10 and 51°C, annual rainfall around 80 mm and annual evaporation around 180 cm (Li et al. 2008). More than 5,000 km^2 of the total Salton Basin are topographically below sea level (Li et al. 2008).

The structural configuration of the area was discussed in numerous studies (i.e., Helgeson 1968; Olmsted et al. 1973; Elders and Biehler 1975; Loeltz et al. 1975; Barragan et al. 2001). According to the above, the Salton Trough which extends inland from the Gulf of California is an area of crustal spreading and a NW–SE directed strike slip plate boundary between the North American and Pacific plates (Fig. 11.25). The trough is an elongated rift system bounded in the northeast by the San Andreas fault system and in the southwest by the San Jacinto fault system. The tectonic movements, which prevail until the present (i.e., Schmitt and Hulen 2008),

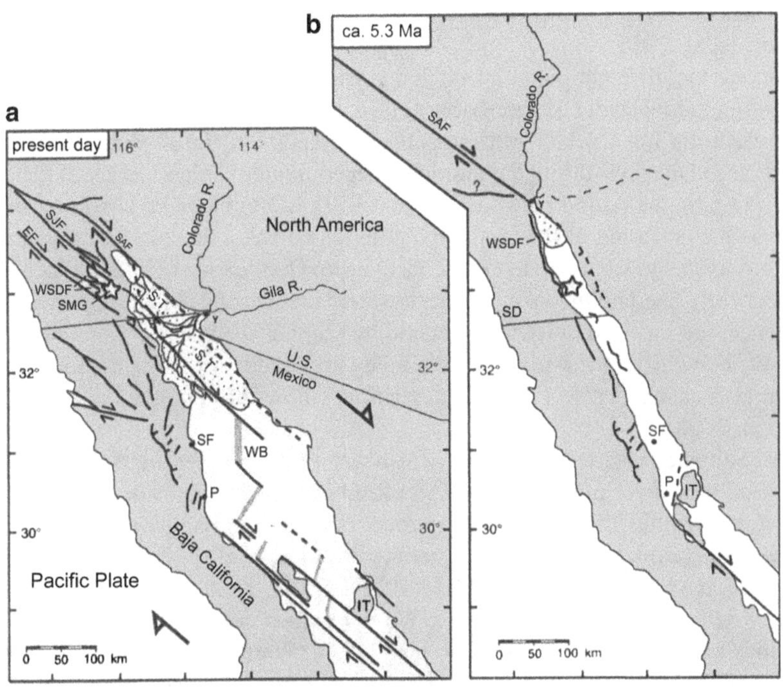

Fig. 11.25 A regional tectonic map of southern California and the Salton Trough [after Dorsey et al. (2007)]. (**a**) Present day situation; (**b**) situation at 5.3 Ma

are responsible for the formation of extensional rifts, pull apart basins and depressions, which are subsiding in places below sea level. Calculated subsidence rates, in the center of the Salton Trough, are between 2.2 and 3.8 mm/y (Schmitt and Hulen 2008). In addition, subsidence rate of up to 12 cm/y, following anthropogenic fluid extraction, was reported from the Cerro Prieto geothermal field (Glowasca et al. 2010).

The motion on the plates' boundary started already in Neogene times (as early as around 12 Ma), resulting in the formation and subsidence of the trough (Dorsey et al. 2007). This, in turn, caused a marine incursion of the basin from the Gulf of California in the south, forming an inland sea arm. The earliest evidence of the marine incursion, in the southern part of the trough, is dated as 8.1 Ma and the subsequent marine flooding of the entire basin occurred in late Miocene to early Pliocene between 6.5 and 6.3 Ma (Oskin and Stock 2003; Dorsey et al. 2007). During the above time span the Salton Trough, assumingly, served as part of the global marine base-level. A subsequent rapid propagation of the Colorado River, from the north to the Salton Trough, is evidenced at the Miocene–Pliocene boundary around 5.33 Ma (Dorsey et al. 2007). Over the past 3 Ma, the Colorado River continuously built a delta which separated the basin from the Gulf of California, yielding multiple lake cycles that occupied the basin (Li et al. 2008). The basin started to accumulate a sequence of evaporites and clastic fill deposits derived from the Colorado River and its delta, as a result, attains a thickness of thousands of meters (Helgeson 1968; Olmsted et al. 1973; Loeltz et al. 1975; Pacheco et al. 2006).

Among the different paleo-lakes, that existed during the above discussed time period, some low level lakes probably served as terminal lakes. Others, such as the paleo-Cahuilla Lake, which existed between 20,500 and 1,300 years BP, experienced closed, or flow-through conditions, depending on their water level (Li et al. 2008). During the period between 700 and 1500 AD the lake level was 12 m asl, whereas the elevation of the topographic divide between the lake and the Gulf of California, was low enough to enable its overflow to the lower Colorado delta and consequently keeping the lake as a fresh water lake. During the entire period of its existence, the lake's salinity was dictated by climate changes, by Colorado River inflows, as well as by evaporation changes and water level fluctuations, all of which also changed it from a closed basin to a flow-through one and vice versa (Li et al. 2008).

The Salton Sea started in 1905 as a fresh water lake. At present its water salinity increased to a level of 41–45 g/l. The salinity is contributed mainly by upward diffusion of brines that dissolve evaporites from the sediments at the bottom of the sea, by fertilizers in irrigation return flows, as well as by evaporation, as described by Helgeson (1968), Rex (1971), Payne et al. (1979), Watts et al. (2001), Schroeder et al. (2002) and Wardlaw and Valentine (2005). The lake salinities are inversely related to the lake levels (Watts et al. 2001). Most of the authors rule out any possibility of mixture with current seawater, based mainly on the chemical and stable isotope affinity of the waters (i.e., Rex 1971; Coplen et al. 1973; Payne et al. 1979).

Groundwater, associated with the hydrothermal fields, that are related to recent rhyolitic volcanism (Schmitt and Hulen 2008) of the Mexicali Valley and Cerro Prieto, south of the Salton Sea, vary in their water type, salinity and isotopic compositions. The shallow phreatic aquifers contain chloride, sulfate and bi-carbonate water types, and salinities can attain levels of 40 g/l related to evaporite dissolution and evaporation (Mckibben and Elders 1985; Portugal et al. 2005). In this matter, it is important to note that Barragan et al. (2001) proposed an additional mechanism of salination, related to the geothermal fields, of convecting seawater along extensional tectonic structures, as was also observed in submarine hydrothermal vents in the nearby Gulf of California. They proposed, accordingly, the mechanism of mixing of fossil seawater and Colorado River waters.

The groundwater flow pattern is convergent into the Salton Sea depression, whereas the basin is separated from the Gulf of California, in the south by a rather shallow groundwater divide, coinciding roughly with the All American Canal and the US-Mexican border (Loeltz et al. 1975). Groundwater flow direction from the divide is to the Salton Depression base level in the north and to the ocean base level in the south (Portugal et al. 2005).

Regarding the salination mechanisms of the Salton Sea waters, and of the adjoining groundwater systems, Kafri (1984) proposed an additional mechanism to those described before, whereby a mixture of relatively fresh water occurs with currently subsurface encroaching seawater from the south. This proposed mechanism is based on the similarity to other places in the world where base-levels below sea level exist close to the sea (see also Sect. 11.2.5). Such a mechanism is supported by the existing combination of the following: (a) A considerable elevation difference that exists between both the marine and the Salton Sea base-levels. (b) A hydrological continuity that, supposedly, exists between both base levels, partly facilitated by the hydraulically conductive trough fill and by the active extensional faults that assumingly act as conduits. (c) The rather low groundwater divide that indeed exists in between both base levels. This proposed mechanism is possibly enhanced by the convecting seawater mechanism described before by Barragan et al. (2001).

11.6 The Chotts of Tunisia and Algeria

The Chotts Trough of Tunisia and Algeria (Fig. 8.1) lies between the Saharan platform in the south and the Atlas Mountains in the north. The region was subjected in the Mesozoic and Cenozoic to tectonic activity and rifting associated with orogeny processes that occurred in the north African-Mediterranean region (Frizon de Lamotte et al. 2009). The trough includes numerous anticlines, synclines, structural blocks and strike slip faults that are still active and resulting in subsidence at different rates (Swezey 1996). The described tectonic activity is responsible to the formation of different depocenters within the trough, that coincide with the different Chotts, and which accumulated sediments since the Miocene (Swezey 1996).

The Chotts (namely, salt pans or sabkhas) of Tunisia and Algeria constitute a series of saline endorheic lakes and salt basins that are occupied in a huge depression or trough, which was formerly an arm of the Mediterranean Sea, extending up to nearly 400 kms west of the Gulf of Gabes in Tunisia (Fig. 11.26). The Chotts are also a remnant of the huge Chott Mega-lake, which existed during different periods, between the Late Pleistocene and the Early Holocene, as suggested by U/Th dating of ancient lake sediments (Richards and Vita-Finzi 1982; Causse et al. 1989). At high water level, the Chotts Mega-lake could overflow across a ca 45 m high divide, located between the lakes and the Mediterranean Sea and flow to the sea, entering it at the Gulf of Gabes.

The huge depression occupies the following Chotts, which act as endorheic base levels, from the Gulf of Gabes in Tunisia in the east to Algeria in the west, all of which are below sea level (bsl): Chott el Fejej, Chott el Djerid (at 16 m bsl), Chott Rharsa (at 21 m bsl), Chott Melrhir (at 36 m bsl) and Chott Merouane (at 40 m bsl).

The major groundwater systems of the region are the following:

(a) The vast Continental Intercalaire or Nubian sandstone aquifer, of Early Cretaceous age, is known from all over the Sahara and North Africa (Edmunds et al. 2003). The aquifer, which is mostly deep seated and confined, was recharged mainly during humid periods in the Quaternary through its exposures in the Sahara, in the south, and the Atlas Mountains in the north. Despite the fact that the aquifer is confined, there are indications that it feeds vertically, in places, the overlying aquifers through faults (i.e., the El Hamma faults near the Gulf of Gabes), or by upward leakage (Kamel et al. 2008; Trabelsi et al. 2009). The

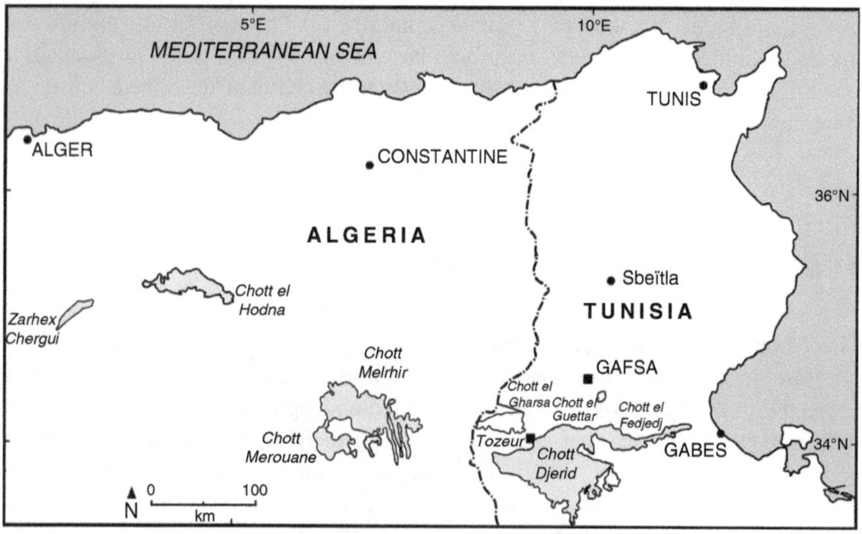

Fig. 11.26 Location map of the Chotts of Algeria and Tunisia

flow path of its paleo-waters converges into the Chotts Troughs and is being discharged upwards through the Chotts area and to the Gulf of Gabes (Fig. 11.27) (Edmunds et al. 2003). Regarding salinity, the waters are usually brackish, gaining their salinity along the flow path through water–rock interaction and dissolution of continental evaporites (Edmunds et al. 2003).

(b) The Complexe Terminal, mostly phreatic aquifer of Miocene to Pliocene age, consists in the Chotts area of a thick, up to 1,300 m, continental detrital sequence with evaporite intercalations (Guendouz et al. 2003). This aquifer also contains paleo waters recharged during humid periods in the late Quaternary and Holocene. Similar to the Continental Intercalaire aquifer, here also

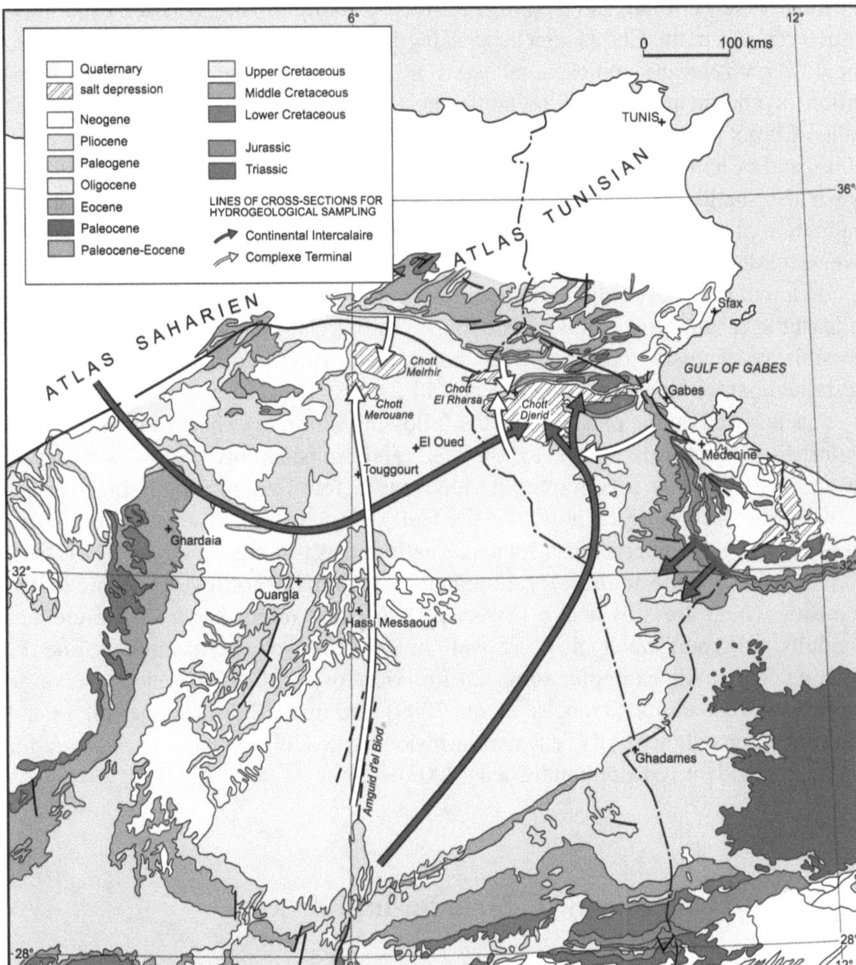

Fig. 11.27 Groundwater flow map of the Continental Intercalaire aquifer showing a convergent flow pattern to the Chotts area [modified after Edmunds et al. (2003)]

groundwater flow converges and being discharged in the Chotts area and to the Gulf of Gabes (Saibi et al. 2009). Groundwater salination takes place along the flow path through evaporite dissolution and anthropogenic contamination (Guendouz et al. 2003, 2006).

(c) The Plio-Pleistocene detrital phreatic aquifer, which attains a thickness of up to 200 m (Kamel et al. 2008). The aquifer is connected to the sea and is known to be intruded by seawater (Gaaloul and Cheng 2003). It is assumed that both latter aquifers are hydraulically connected to the sea in the Gulf of Gabes (see below).

The Chotts of Tunisia and Algeria, thus, serve as terminal endorheic base-levels to convergent groundwater flow of the main, namely the Continental Intercalaire, the Complexe Terminal and the Plio-Pleistocene aquifers. The evolution of salinity, which can rise in the Chotts area up to 350 g/l (Zammouri et al. 2007), according to the different relevant studies, is related to mineralization of groundwater by dissolution of continental evaporites along the flow path, by evaporation within the saline Chotts that serve as discharge zones (Edmunds et al. 2003; Guendouz et al. 2003) and by anthropogenic sources such as by domestic pollution and by fertilizers when the shallow phreatic aquifers are regarded (Guendouz et al. 2003, 2006). Salination of coastal aquifers of the Gulf of Gabes was also suggested as a result of over-exploitation and seawater encroachment (Gaaloul and Cheng 2003).

On top of the above salination factors, current subsurface seawater intrusion into at least the easternmost Chotts, close to the Gulf of Gabes, is suggested herein as a possible mechanism that should be considered with accordance to the model described before (see also Sects. 6.5 and 11.2).

This is based on the existence of the following setup: (a) The eastern Chotts are endorheic base-levels close to the sea and considerably below sea level. (b) A relatively low topographic divide, only a few tens of meters high, exists between the easternmost Chotts and the Gulf of Gabes, possibly permitting intrusion of seawater underneath a relative shallow fresh–saline water interface, still above the base of the aquifer. (c) Abundant east–west and northwest striking faults, some of which are still active (Swezey 1996), that might serve as preferential conduits to groundwater flow as well as to seawater encroachment. The El Hamma faults, for example, some 20 km west of the Gulf of Gabes, serve as groundwater conduits (Trabelsi et al. 2009) and may thus also permit inland seawater encroachment. (d) Seawater intrusion to the Gulf of Gabes coastal aquifer is already evident (Gaaloul and Cheng 2003).

11.7 Sabkhat Ghuzayyil, Sirte Basin, Libya

Sabkhat Ghuzayyil (Fig. 8.1) is an endorheic landlocked basin, some 47 m below sea level, located in the northern part of the Sirte Basin, at coordinates 29°N 19°E, some 60 kms south of the embayment of the Gulf of Sirte (Fig. 11.28).

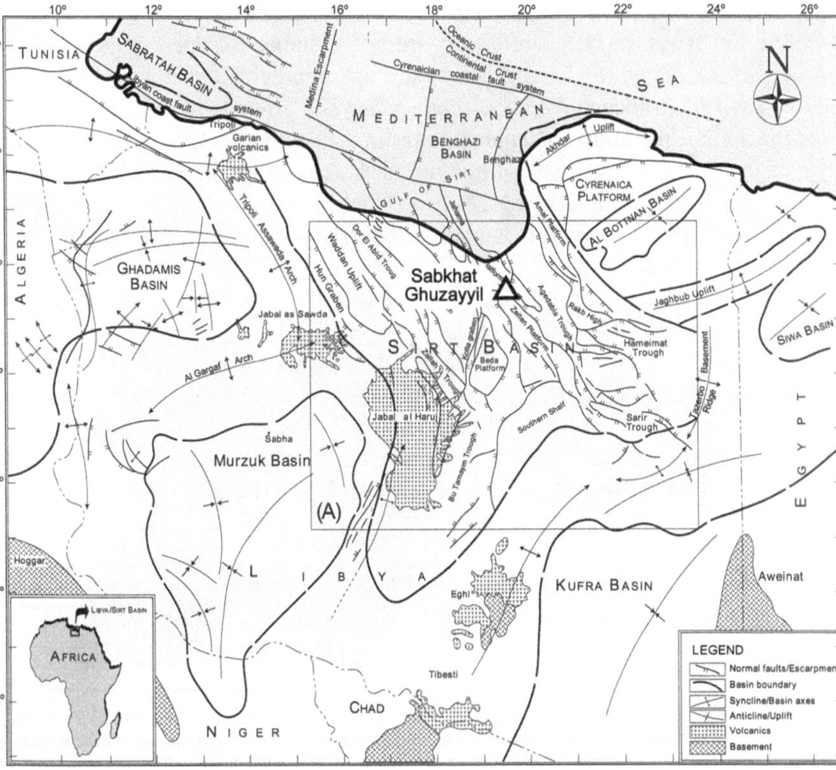

Fig. 11.28 The structural framework of the Sirte Basin, Libya [after Abadi et al. (2008)]. Also shown is the location of Sabkhat Ghuzayyil

Tectonically speaking, the entire area consists of a huge tectonic trough system, that extends from the Kufra Basin in southeast Libya to the Sirte Basin and the Gulf of Sirte in the northwest (Fig. 11.28). The trough was subjected to consecutive phases of faulting, rifting, subsidence and fault reactivation since the Jurassic and until the present, related to stretching and extensional processes, tied to the opening of the Atlantic and the Tethys. As a result, the trough is dissected by several faults forming blocks, horsts and grabens, mostly in the northwest direction. The subsiding trough has accumulated thousands of meters thick sediments (Van Houten 1983; Gumati and Nairn 2007; Abadi et al. 2008). In addition, this huge trough occupied since the middle Miocene a regional paleo river system, 900 kms long, that connected the Kufra Basin through the Sirte Basin to the Mediterranean Sea (Paillou et al. 2009).

The hydrogeology of the entire region was described in detail by Edmunds and Wright (1979), Wright et al. (1982) and Lloyd (1990). A concise summary of the hydrogeological setup is given below:

The main regional groundwater systems in the area are the Nubian Sandstone aquifer (or system) and the Post Eocene aquifer (or Tertiary system).The combined thickness of both aquifers is some 1,000 m. The Nubian Sandstone continental aquifer is known mainly in the south in the Kufra Basin where it is phreatic and contains fresh water. The aquifer, according to stable isotope composition and radiogenic dating, was recharged mainly during humid climates in Late Quaternary (35–15 ka) and Holocene (8–5 ka) time spans. The groundwater flow direction is from the intake area in the Sudan mountains in the south, diverging to the discharge zone of the Qattara Depression in the northeast, and to the Sirte Basin depression and sabkhas, which are also below sea level (Fig. 11.29). The Post Eocene aquifer is known from the Sirte Basin. It is partly recharged by current direct recharge and partly from the underlying Nubian Sandstone aquifer. The groundwater flow here, also, is northward to the coastal aquifers and the discharge zone of Sabkhat Ghuzayyil where groundwater level is below sea level.

Salinities of groundwater increase along the flow paths, as a result of water rock interaction and evaporation of shallow groundwater, from fresh water in the south to water salinity as high as 100 g/l in the north. A freshening, in places of the Post Eocene aquifer, is related to vertical recharge from the superimposed paleo-river system described above (Edmunds and Wright 1979; Wright et al. 1982).

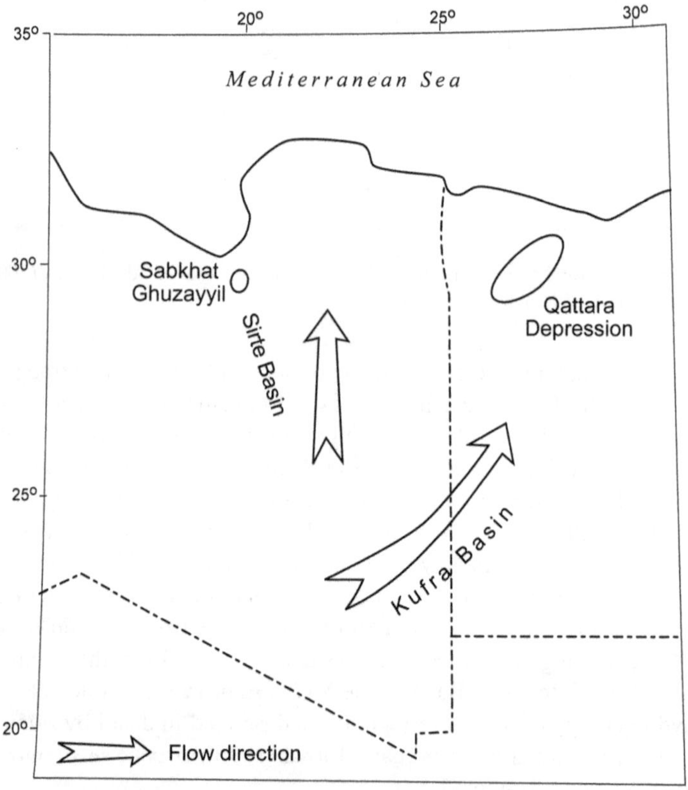

Fig. 11.29 Map showing the main groundwater flow directions in the Kufra and Sirte basins, Libya [modified after Wright et al. (1982)]. Permission to use this figure was granted by the Geological Society of London

It is proposed to consider the possibility that, on top of the groundwater salination mechanisms mentioned above, an additional mechanism of current subsurface seawater encroachment to this base level below sea level, takes place. This is based on the prevailing combination, as described from other localities in this chapter, of an existing base level below sea level that is possibly hydraulically connected to the sea, coupled with a hydraulically conductive fault system.

11.8 Lago Enriquillo, Dominican Republic

Lago Enriquillo, in the Dominican Republic, at coordinates 18°N 71°W, is a hypersaline lake, at an elevation of some 46 m bsl, situated a few tens of kilometers west of the Caribbean Sea (Figs. 8.1 and 11.30). The lake is situated within a 85 km long and 12 km wide morpho-tectonic depression or a rift valley, named the Cul de Sac depression which extends from the Dominican Neiba embayment on the Caribbean Sea to Haiti in the west (Mann et al. 1984; Chiesa and Mazzoleni 2001). The area, which is situated within the boundary zone between the Caribbean plate and the North American plate, is transected by abundant, still active, strike slip and reverse faults, which are responsible for the current seismic activity in the area (Mann et al. 1984; Tuttle et al. 2003; DeMets and Wiggins-Grandison 2007).

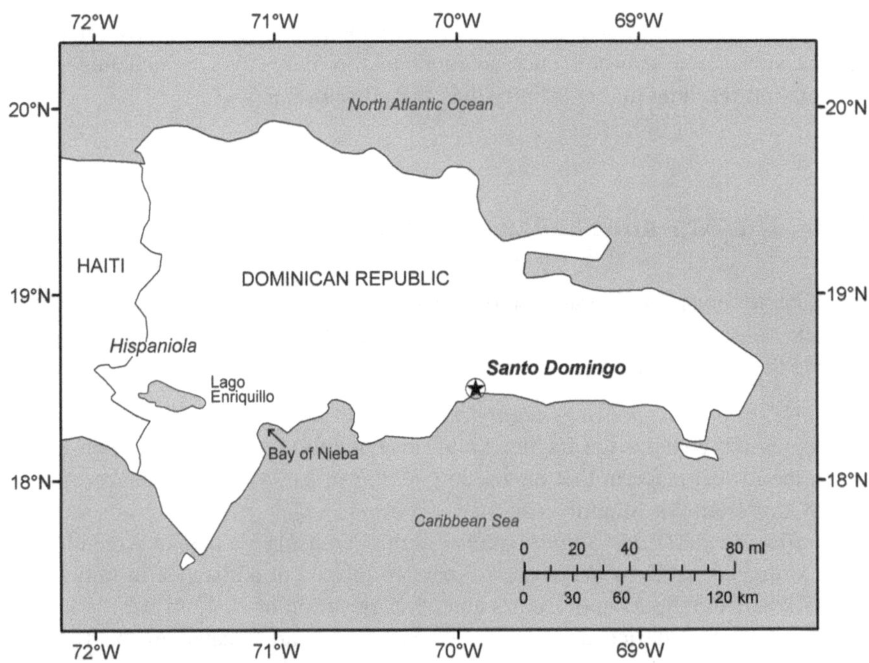

Fig. 11.30 Location map of Lago Enriquillo, Dominican Republic

At the peak of the last glacial period, some 18,000 years ago, the Enriquillo depression was above sea level draining to the considerably low, at 120 m bsl, global and Caribbean Sea base-level through the present day Neiba embayment (Mann et al. 1984). The subsequent rising sea level in early Holocene time, at ca 9.8 ka, flooded the Enriquillo depression, creating an inland sea arm, which existed until ca 4,900 ka. At around 4,900 ka, the marine sea arm was disconnected from the sea, as a result of the combination of damming of the eastern mouth of the valley by deltaic deposits, and possible due to vertical movements and of climate change. Following the restriction of the embayment, it became a closed hypersaline lagoon and its level dropped to the present one, below sea level, due to the high evaporation of this arid area and the continuous subsidence of the depression since then (Mann et al. 1984; Doss et al. 2005; Medley 2006; Winsor et al. 2006).

The present total salinity, as well as the different ion concentrations and stable isotope composition of the lake water, were described and discussed in detail by Buck et al. (2005). The hypersaline nature of the water is related to evaporation as also is exhibited by the stable isotope composition. The total salinity of the lake water fluctuates annually and seasonally in response to recharge and evaporation. Historic salinities ranged from about 35 g/l (normal marine salinity) to over 100 g/l. At present, the total seasonal salinity range is between ca 80 and 105 g/l with a chlorine concentration factor of 2.25 as compared to Caribbean seawater. It is interesting to note that the initial marine signature of the waters is not obscured by the evaporation process, as shown by the calculated Na/Cl equivalent ratio, around 0.87, resembling the Caribbean seawater.

Here, too, it is proposed that the saline end member could be partly related to current subsurface seawater encroachment to this base level, which underwent evaporation reaching higher salinity than that of seawater.

11.9 The Argentine Salinas Close to the Atlantic Coast

A few depressions, considerably below sea level, are located along, and close to, the Atlantic coast (Fig. 8.1). The depressions occupy lagoons, salinas and salt pans. Among the known ones are:

(a) *The Gran Bajo de San Julian depression*, at coordinates 49°S 68°W. The depression contains the Laguna Guadalosa and the Laguna Del Carbon which is the lowest place in that depression, at 105 m bsl, and at a distance of only 18 kms from the Atlantic coast.
(b) *Salinas Grandes and Salinas Chicas* in the Peninsula Valdes, in Argentine, at coordinates 42°S 64°W. These salinas are located at a distance of only some 17 kms from the Atlantic coasts and at an elevation of ca 42 m bsl.
(c) The Laguna Chasico and Salinas Chicas which are located in the Chasico Depression at coordinates 38°S 62°W. The depression is close to the Atlantic coast, bounded by faults, and is situated at elevations between 20 and 40 m bsl.

Despite the scarce available information about the hydrogeological setup and the salination processes of the above depressions that occupy salinas, it can be speculated that they resemble similar base-levels below sea level and close to the sea, regarding the hydrological processes that are responsible to salination processes, as described in this chapter of the book.

11.10 Sebkha Paki Tah, Morocco

Sabkha Paki Tah, in Morocco (Fig. 8.1) is a depression that is occupied by a sabkha, located at coordinates 27°N 12°W. The distance of the depression from the Atlantic coast is only some 11 kms and it attains an elevation of 55 m bsl.

It is assumed, that its hydrological setup resemble other similar base-levels below sea level described above.

11.11 Salination Mechanism

As mentioned before, the above described base-levels occupy saline lakes, sabkhas or salinas and are hydraulically connected to adjoining saline groundwater systems. The typical high salinities of these base-levels can be attributed to the location of a terminal base-level in an arid or semi-arid zone where evaporation exceeds the inflow to the basin, combined with the existence of trapped saline waters and/or dissolution of previously deposited evaporites.

An additional salination mechanism is suggested, regarding those specific base-levels that are considerably below sea level and close to the sea (Kafri and Arad 1979; Kafri 1984; Kafri et al. 2007a, b) (see also Sect. 11.2.5). This mechanism suggests a current subsurface seawater encroachment from the upper marine base-level to the nearby lower continental closed base-level, facilitated by the elevation difference between both, and provided that a hydrological continuity exists between those base-levels. Fault systems, very often active, can serve as conduits to groundwater flow as well as to the inland subsurface seawater flows. In addition, the low groundwater divide between both base-levels results in a shallow fresh/saline water interface above the base of the aquifer which, in turn, also permits this subsurface seawater encroachment into the lower base-level across the divide. It should be mentioned, that Nunn and Harris (2007) proposed a somewhat similar model of subsurface seepage of seawater across a barrier, concerning subsurface seawater intrusion from the Atlantic Ocean to the declining paleo Messinian Mediterranean base-level (see also Sect. 7.2).

It can be assumed that the rate of salination of such base-levels can be affected by the changing rates and amounts of the inland seawater inflows, following changes of the base-levels elevations, and the resultant elevation differences between them and depending of the hydraulic properties of media in between.

References

Abadi AM, van Wees J-D, van Dijk PM, Cloetingh SAPL (2008) Tectonics and subsidence evolution of the Sirte Basin, Libya. Am Assoc Pet Geol Bull 92:993–1027

Abbo H, Shavit U, Markel D, Rimmer A (2003) A numerical study on the influence of fractured regions on lake/groundwater interaction; the Lake Kinneret (Sea of Galilee) case. J Hydrol 283:225–243

Albritton Jr CC, Brooks JE, Issawi B, Swedan A (1990) Origin of the Qattara depression, Egypt. Geol Soc Am Bull 102:952–960

Allard P, Tazieff H, Dajlevic D (1979) Observations of seafloor spreading in Afar during the November 1978 fissure eruption. Nature 279:30–33

Arad A, Kafri U, Fleisher E (1975) The Na'aman springs, northern Israel: salination mechanism of an irregular freshwater–seawater interface. J Hydrol 25:81–104

Arad A, Michaeli A (1967) Hydrogeological investigations in the western catchment of the Dead Sea. Isr J Earth Sci 16:181–196

Aref MAM, El-Khoriby E, Hamdan MA (2002) The role of salt weathering in the origin of the Qattara Depression, Western Desert, Egypt. Geomorphology 45:181–195

Audin L, Manighetti I, Tapponnier P, Metivier F, Jacques E, Huchon P (2001) Fault propagation and climatic control of sedimentation on the Ghoubbet rift Floor: insights from the Tadjouraden cruise in the western Gulf of Aden. Geophys J Int 144:1–28

Ayele A, Nyblade AA, Langston CA, Cara M, Leveque J-J (2006) New evidence for Afro-Arabian plate separation in southern Afar. In: Yirgu G, Ebinger CG, Maguire PKH (eds) The Afar volcanic province within the East African rift system. Geological Society of London Special Publications, London 259:133–141

Baer G, Schattner U, Wachs D, Sandwell D, Wdowinski S, Frydman S (2002) The lowest place on earth is subsiding – an InSAR (interferometric synthetic aperture radar) perspective. Geol Soc Am Bull 114:12–23

Barberi F, Ferrara G, Santacroce R, Varet J (1975) Structural evolution of the Afar triple junction. In: Pilger A, Rosler A (eds) Afar depression of Ethiopia. Schweitzersche Verlag, Stuttgart 38–54

Barragan RRM, Birkle P, Portugal ME, Arellano GVM, Alvarez R (2001) Geochemical survey of medium temperature geothermal resources from the Baja California Peninsula and Sonora, Mexico. J Volcanol Geotherm Res 110:101–119

Bartov Y, Agnon A, Enzel Y, Stein M (2006) Late quaternary faulting and subsidence in the central Dead Sea basin. Isr J Earth Sci 55:17–31

Bartov Y, Sagy A (2004) Late Pleistocene extension and strike-slip in the Dead Sea. Geol Mag 141:565–572

Batayneh AT (2006) Use of electrical resistivity methods for detecting subsurface fresh and saline water and delineation their interfacial configuration: a case study of the eastern Dead Sea coastal aquifers, Jordan. Hydrogeol J 14:1277–1283

Begin ZB (2002) Computer simulation of salinity in the Late Quaternary Lake Lisan. Isr J Earth Sci 51:225–232

Begin ZB, Stein M, Katz A, Machlus M, Rosenfeld A, Buchbinder B, Bartov Y (2004) Southward migration of rain tracks during the last glacial, revealed by salinity gradient in Lake Lisan (Dead Sea rift). Quat Sci Rev 23:1627–1636

Begin ZB, Zilberman E (1997) Main stages and rate of relief development in Israel. Geol Surv Isr Rep GSI/24/97 (In Hebrew)

Ben Menahem A (1981) Variation of slip and creep along the Levant Rift over the past 4500 years. Tectonophysics 80:183–197

Bergelson G, Nativ R, Bein A (1999) Salinization and dilution history of ground water discharging into the Sea of Galilee, the Dead Sea Transform, Israel. Appl Geochem 14:91–118

Bookman (Ken-Tor) R, Enzel Y, Agnon A, Stein M (2004) Late Holocene lake levels of the Dead Sea. Geol Soc Am Bull 116:555–571

Bowman D (1971) Geomorphology of the shore terraces of the late Pleistocene Lisan Lake (Israel). Palaeogeogr Palaeoclimatol Palaeoecol 9:183–209

Buck DG, Brenner M, Hodell DA, Curtis JH, Martin JB, Pagani M (2005) Physical and chemical properties of hypersaline Lago Enriquillo, Dominican Republic. Verh Internat Verein Limnol 29:1–7

Buck WR (2006) The role of magma in the development of the Afro-Arabian rift syste. In: Yirgu G, Ebinger CG, Maguire PKH (eds) The Afar volcanic province within the East African rift system. Geological Society of London Special Publications 259:43–54

Causse C, Coque R, Fontes J Ch, Gasse F, Gilbert E, Ben Quezdou H, Zouari K (1989) Two high levels of continental waters in the southern Tunisian chotts at about 90 and 150 ka. Geology 17:922–925

Chiesa S, Mazzoleni G (2001) Dominican Republic (Hispaniola Island, north-eastern Caribbean): a map of morpho-structural units at a scale 1:500,000 through LANDSAT TM image interpretation. Rev Geol Am Cent 25:99–106

Coplen TB, Combs J, Elders WA, Rex RW, Burckhalter G, Laird R (1973) Preliminary findings of an investigation of the dunes thermal anomaly, Imperial Valley, California. State Calif Dept Water Res 48 pp

DeMets C, Wiggins-Grandison M (2007) Deformation of Jamaika and motion of the Gonave microplate from GPS and seismic data. Geophys J Int 168:362–378

Dorsey RJ, Fluette A, McDougall K, Housen BA, Janecke SU, Axen GJ, Shirvell CR (2007) Chronology of Miocene–Pliocene deposits at split mountain Gorge, Southern California: a record of regional tectonics and Colorado River evolution. Geology 35:57–60

Doss WC, Greer L, Curran HA, Patterson WP, Mortlock RA (2005) Large tufa-coated serpulid mounds signal and abrupt mid-Holocene transition from marine to restricted hyposaline conditions, Lago Enriquillo, Dominican Republic. Geol Soc Am Abstr Program 37:365

Doubre C, Peltzer G (2007) Fluid-controlled faulting process in the Asal Rift, Djibouti, from 8 yr of radar interferometry observations. Geology 35:69–72

Edmunds WM, Guendouz AH, Mamou A, Moulla A, Shand P, Zouari K (2003) Groundwater evolution in the Continental Intercalaire aquifer of southern Algeria and Tunisia: trace element and isotopic indicators. Appl Geochem 18:805–822

Edmunds WM, Wright EP (1979) Groundwater recharge and palaeoclimate in the Sirte and Kufra basins, Libya. J Hydrol 40:215–241

El Shazly EM, Abdel Hady MA, Ghawaby MAE, Khawasik SM, El Shazly MM (1976) Geologic interpretation of Landsat Satellite images for the Qattara Depression area, Egypt. Egypt OSU Remote Sensing Center, Cairo 54

Elders WA, Biehler S (1975) Gulf of California rift system and its implications for the tectonics of western North America. Geology 3:85–87

Froehlich K, Aggarwal PK, Garner WA (2007) An integrated approach in evaluating isotope dating of the Nubian Sandstone Aquifer System (NSAS) in Egypt. In: Advances in isotope hydrology and its role in sustainable water resources management (HIS-2007) Proc Symp 1:31–45

Faure H, Walter RC, Grant DR (2002) The coastal oasis: ice age springs on emerged continental shelves. Global Planet Change 33:47–56

Fouillac AM, Fouillac C, Cesbron F, Pillard F, Legendre O (1989) Water–rock interaction between basalt and high-salinity fluids in the Asal Rift, Republic of Djibouti. Chem Geol 76:271–289

Frizon de Lamotte DF, Leturmy P, Missenard Y, Khomsi S, Ruiz G, Saddiqi O, Guillocheau F, Michard A (2009) Mesozoic and Cenozoic vertical movements in the Atlas system (Algeria, Morocco, Tunisia): an overview. Tectonophysics 475:9–28

Frumkin A (2001) The cave of the letters sediments-indication of an early phase of the Dead Sea depression? J Geol 109:79–90

Frumkin A, Fischhendler I (2005) Morphometry and distribution of isolated caves as a guide for phreatic and confined paleohydrological conditions. Geomorphology 67:457–471

Gaaloul N, Cheng AH-D (2003) Hydrogeological and hydrochemical investigation of coastal aquifers in Tunisia-Crisis in overexploitation and salinization. Second International Conference on Saltwater Intrusion and Coastal Aquifers, Merida, Mexico, March 30-April 2, 2003

Garfunkel Z (1981) Internal structure of the Dead Sea leaky transform (rift) in relation to plate kinematics. Tectonophysics 80:81–108

Garfunkel Z, Ben-Avraham Z (1996) The structure of the Dead Sea basin. Tectonophysics 266:155–176

Garfunkel Z, Beyth M (2006) Constraints on the structural development of Afar imposed by the kinematics of the major surrounding plates. In: Yirgu G, Ebinger CG, Maguire PKH (eds) The Afar volcanic province within the East African rift system. Geological Society of London Special Publication, London 259:23–42

Garfunkel Z, Zak Y, Freund R (1981) Active faulting in the Dead Sea rift. Tectonophysics 80:1–26

Gasse F, Rognon P, Street FA (1980) Quaternary history of the Afar and Ethiopian Rift lakes. In: Williams MAJ and Faure H (eds) The Sahara and the Nile. Quaternary environments and prehistoric occupation in northern Africa, Balkema, Rotterdam 361–400

Gasse F, Fontes J-C (1989) Palaeoenvironments and palaeohydrology of a tropical closed lake (Lake Asal, Djibouti) since 10,000 yr B.P. Palaeogeogr Palaeoclimatol Palaeoecol 69:67–102

Glowasca E, Sarychikhina O, Suarez F, Nava FA, Mellors R (2010) Anthropogenic subsidence in the Mexicali Valley, Baja California, Mexico, and slip on the Saltillo fault. Environ Earth Sci 59:1515–1524

Goldshmidt MJ, Arad A, Neev D (1967) The mechanism of the saline springs in the Lake Tiberias depression. Geol Surv Isr Bull 45:1–19

Gonfiantini R, Borsi S, Ferrara G, Panichi C (1973) Isotopic composition of waters from the Danakil depression (Ethiopia). Earth Planet Sci Lett 18:13–21

Gossel W, Ebraheem AM, Wycisk P (2004) Avery large scale GIS-based groundwater flow model for the Nubian sandstone aquifer in Eastern Sahara (Egypt, northern Sudan and eastern Libya) Hydrogeol J 12:698–713

Guendouz A, Moulla AS, Edmunds WM, Zouari K, Mamou A (2003) Hydrogeochemical and isotopic evolution of water in the Complexe Terminal aquifer in the Algerian Sahara. Hydrogeol J 11:483–495

Guendouz A, Moulla AS, Remini B, Michelot JL (2006) Hydrochemical and isotopic behaviour of a Saharan phreatic aquifer suffering severe natural and anthropic constrains (case of Oued-Souf region, Algeria). Hydrogeol J 14:955–968

Gumati YD, Nairn AEM (2007) Tectonic subsidence of the Sirte Basin, Libya. J Petrol Geol 14:93–102

Gvirtzman H (2006) Groundwater hydrology and paleohydrology of the Dead Sea rift valley. In: Enzel Y, Agnon A, Stein M (eds) New frontiers in Dead Sea paleoenvironmental research. Geological Society of America Special Paper, Boulder, CO 401:95–111

Gvirtzman H, Garven G, Gvirtzman G (1997) Hydrogeological modeling of the saline hot springs at the Sea of Galilee, Israel. Water Resour Res 33:913–926

Heimann A, Steinitz G (1989) 40Ar/39Ar total gas ages of basalts from Notera #3 well, Hula Valley, Dead Sea Rift: stratigraphic and tectonic implications. Isr J Earth Sci 38:173–184

Heinl M, Brinkmann P (1989) A groundwater model of the Nubian aquifer system. J Sci Hydrol 34:425–447

Helgeson HC (1968) Geologic and thermodynamic characteristics of the Salton Sea geothermal system. Am J Sci 266:129–166

Horowitz A (2001) The Jordan Rift Valley. Balkema, Lisse, The Netherlands, 730

Houssein I, Jalludin M (1996) The salinity of Djibouti's aquifer. J Afr Earth Sci 22:409–414

Hurwitz S, Stanislavsky E, Lyakhovsky V, Gvirtzman H (2000) Transient groundwater–lake interactions in a continental rift: sea of Galilee, Israel. Geol Soc Am Bull 112:1694–1702

Issar A, Bein A, Michaeli A (1972) On the ancient water of the upper Nubian Sandstone aquifer in central Sinai and southern Israel. J Hydrol 17:353–374

Kafri U (1967) Facies changes in the southwestern Carmel (Israel) and their influence on groundwater regime. Isr J Earth Sci 16:206–214

Kafri U (1969) Geology and groundwater of the Cenomanian formations in Galilee, west of the watershed. PhD Thesis, Hebrew University, Jerusalem, 188 (in Hebrew)

Kafri U (1982) Relationship between Dead Sea and groundwater levels in the western Dead Sea catchment area. Isr Geol Surv Curr Res 1981:90–94

Kafri U (1984) Current subsurface seawater intrusion to base levels below sea level. Environ Geol Water Sci 6:223–227

Kafri U, Arad A (1978) Paleohydrology and migration of the ground-water divide in regions of tectonic instability in Israel. Geol Soc Am Bull 89:1723–1732

Kafri U, Arad A (1979) Current subsurface intrusion of Mediterranean seawater – a possible source of groundwater salinity in the Rift Valley System, Israel. J Hydrol 44:267–287

Kafri U, Ecker A (1964) Neogene and quaternary subsurface geology and hydrogeology of the Zevulun Plain. Isr Geol Surv Bull 37:1–13

Kafri U, Goldman M, Lang B (1997) Detection of subsurface brines, freshwater bodies and the interface configuration in-between by the time domain electromagnetic method in the Dead Sea Rift, Israel. Environ Geol 31:42–49

Kafri U, Goldman M, Levi E (2008) The relationship between saline groundwater within the Arava Rift Valley in Israel and the present and ancient base levels as detected by deep geoelectromagnetic soundings. Environ Geol 54: 1435–1445

Kafri U, Goldman M, Lyakhovsky V, Scholl C, Helwig S, Tezkan B (2007a) The configuration of the fresh–saline groundwater interface in northern Israel between the Mediterranean and the Dead Sea base levels as delineated by deep geoelectromagnetic soundings. Geol Surv Isr Geophys Inst Isr Int Rep 25 pp

Kafri U, Goldman M, Lyakhovsky V, Scholl C, Helwig S, Tezkan B (2007b). The configuration of the fresh–saline groundwater interface within the regional Judea Group carbonate aquifer in northern Israel between the Mediterranean and the Dead Sea base levels as delineated by deep geoelectromagnetic soundings. J Hydrol 344:123–134

Kafri U, Horowitz A (2003) Possible northward drainage of the Pliocene Hula Valley. Isr J Earth Sci 52:185–190

Kafri U, Kaufman A, Magaritz M (1983) The rate of Pleistocene subsidence and sedimentation in the Hula Basin as compared with those of other time spans in other Israeli tectonic regions. Earth Planet Sci Lett 65:126–132

Kafri U, Lang B (1987) New data on the young Quaternary fill of the Hula Basin, Israel. Isr J Earth Sci 36:73–82

Kafri U, Yechieli Y (2010) The role of hydrogeological base level in the formation of sub-horizontal caves horizons, example from the Dead Sea Basin, Israel. Environ Earth Sci DOI:10.1007/s12665-009-0435-4

Kamel S, Younes H, Chkir N, Zouari K (2008) The hydro geochemical characterization of ground water in Tunisian Chott's region. Environ Geol 54:843–854

Katz A, Kolodny Y, Nissenbaum A (1977) The geochemical evolution of the Pleistocene Lake Lisan-Dead Sea system. Geochim Cosmochim Acta 41:1609–1626

Katz A, Starinsky A (2009) Geochemical History of the Dead Sea. Aquat Geochem 15:159–194

Ken-Tor R, Agnon A, Enzel Y, Stein M, Marco S, Negendank JFW (2001) High-resolution geological record of historic earthquakes in the Dead Sea basin. J Geophys Res 106:2221–2234

Kiro Y (2006) The effect of the Dead Sea level drop in the past 50 years on the groundwater system in the alluvial aquifer in its vicinity. M Sc Thesis, Hebrew University of Jerusalem, Israel, 116 pp (in Hebrew)

Kiro Y, Yechieli Y, Shalev E, Lyakhovsky V, Starinsky A (2008) Time response of the water table and saltwater transition zone to a base level drop. Water Resour Res DOI:10.1029/2007WR006752

Klein-BenDavid O, Sass E, Katz A (2004) The evolution of marine evaporitic brines in inland basins: the Jordan-Dead Sea Rift valley. Geochim Cosmochim Acta 68:1763–1775

Lensky NG, Dvorkin Y, Lyakhovsky V, Gertman I, Gavrieli I (2005) Water, salt and energy balances of the Dead Sea. Water Resour Res 41:1–13
Levitte D, Eckstein Y (1979) Correlation between the silica concentration and the orifice temperature in the warm springs along the Jordan-Dead Sea Rift Valley. Geothermics 7:1–8
Li H-C, Xu X-M, Ku T-L, You C-F, Buchheim HP, Peters R (2008) Isotopic and geochemical evidence of palaeoclimate changes in Salton Basin, California, during the past 20 kyr: 1. $\delta^{18}O$ and $\delta^{13}C$ records in Lake Tufa deposits. Palaeogeogr Palaeoclimatol Palaeoecol 259:182–197
Lisker S, Vaks A, Bar-Matthews M, Porat R, Frumkin A (2009) Stromatolites in caves of the Dead Sea Fault Escarpment: implications to latest Pleistocene lake levels and tectonic subsidence. Quat Sci Rev 28:80–92
Loeltz OJ, Irelna B, Robison JH, Olmsted FH (1975) Geohydrologic reconnaissance of the Imperial Valley, California. U S Geol Surv Prof Pap 486-K, 54
Lloyd JW (1990) Groundwater resources development in the eastern Sahara. J Hydrol 119:71–87
Mann P, Taylor FW, Burke K, Kulstad R (1984) Subaerially exposed Holocene coral reef, Enriquillo Valley, Dominican Republic. Geol Soc Am Bull 95:1084–1092
Manighetti I, Tapponnier P, Courtillot V, Gallet Y (2001) Strain transfer between disconnected, propagating rifts in Afar. J Geophys Res 106:13613–13665
Marco S, Stein M, Agnon A, Ron H (1996) Long-term earthquake clustering: a 50,000-year paleoseismic record in the Dead Sea Graben. J Geophys Res 101:6179–6191
Marcus E, Slager J (1985) The sedimentary-magmatic sequence of the Zemach 1 well (Jordan-Dead Sea rift, Israel) and its emplacement in time and space. Isr J Earth Sci 34:1–10
Mazor E, Mero F (1969a) The origin of the Tiberias-No'it mineral water association in the Tiberias-Dead Sea Rift Valley, Israel. J Hydrol 7:318–333
Mazor E, Mero F (1969b) Geochemical tracing of mineral and fresh water sources in the Lake Tiberias basin, Israel. J Hydrol 7:276–317
McKibben MA, Elders WA (1985) Fe–Zn–Cu–Pb mineralization in the Salton Sea geothermal system, Imperial Valley, California. Econ Geol 80:539–559
Medley PR (2006) Paleohydrology of middle Holocene lagoonal & lacustrine deposits in the Enriquillo Valley, Dominican Republic: pore morphometrics and isotope geochemistry of ostracoda. Geol Soc Am Abstr Program 38:444
Migowski C, Agnon A, Bookman R, Negendank JFW, Stein M (2004) Recurrence pattern of Holocene earthquakes along the Dead Sea transform revealed by varve-counting and radiocarbon dating of lacustrine sediments. Earth Planet Sci Lett 222:301–314
Miller KG, Kominz MA, Browning JV, Wright JD, Mountain GS, Katz ME, Sugarman PJ, Cramer BS, Christe-Blick N, Pekar SF (2005) The Phanerozoic record of global sea-level change. Science 310:1293–1298
Mlynarski M, Zlotnicki J (2001) Fluid circulation in the active emerged Asal rift (east Africa, Djibouti) inferred from self-potential and Telluric–Telluric prospecting. Tectonophysics 339:455–472
Neev D, Emery KO (1967) The Dead Sea, depositional processes and environments of the evaporates. Geol Surv Isr Bull 41:1–147
Nunn JA and Harris NB (2007) Subsurface seepage of seawater across a barrier: a source of water and salt to peripheral salt basins. Geol Soc Am Bull 119:1201–1217
Olmsted FH, Loeltz OJ, Irelan B (1973) Geohydrology of the Yuma area, Arizona and California. U S Geol Surv Prof Pap 486-H
Oskin M, Stock J (2003) Marine incursion synchronous with plate-boundary localization in the Gulf of California. Geology 31:23–26
Pacheco M, Martin-Barajas A, Elders W, Espinosa-Cardena JM, Helenes J, Segura A (2006) Stratigraphy and structure of the Altar basin of NW Sonora: implications for the history of the Colorado River delta and the Salton trough. Rev Mex Cienc Geol 23:1–22
Paillou P, Schuster M, Toth S, Farr T, Rosenqvist A, Lopez S, Malezieux J-M (2009) Mapping of a major paleodrainage system in eastern Libya using orbital imaging radar: the Kufrah River. Earth Planet Sci Lett 277:327–333

Patterson LJ, Sturchio NC, Kennedy BM, van Soet MC, Sultan M, Lu Z-T, Lehmann B, Purtschert R, El Alfy Z, El Kaliouby B, Dawood Y, Abdallah A (2005) Cosmogenic, radiogenic, and stable isotopic constraints on groundwater residence time in the Nubian Aquifer, Western Desert of Egypt. Geochem Geophys Geosyst 6:1–19

Payne BR, Quijano L, Latorre DC (1979) Environmental isotopes in a study of the origin of salinity of groundwater in the Mexicali Valley. J Hydrol 41:201–215

Pizzi A, Coltorti M, Abebe B, Disperati L, Sacchi G, Salvini R (2006) The Wonji fault belt (Main Ethiopian Rift): structural and geomorphological constraints and GPS monitoring. In: Yirgu G, Ebinger CG, Maguire PKH (eds) The Afar volcanic province within the East African Rift System. Geological Society of London Special Publications 259:191–207

Portugal E, Izquierdo G, Truesdell A, Alvarez J (2005) The geochemistry and isotope hydrology of the Southern Mexicali Valley in the area of the Cerro Prieto, Baja California (Mexico) geothermal field. J Hydrol 313:132–148

Rex RW (1971) Cooperative geological–geophysical–geochemical investigations of geothermal resources in the Imperial Valley area of California. University of California, Riverside, CA, 1–153

Richards GW, Vita-Finzi C (1982) Marine deposit 35,000-25,000 years old in the Chott Djerid, southern Tunisia. Nature 295:54–55

Rimmer A, Gal G (2003) Estimating the saline springs component in the solute and water balance of Lake Kinneret, Israel. J Hydrol 284:228–243

Saibi H, Semmar A, Mesbah M, Ehara S (2009) Variographic analysis of water table data from the Oued-Souf phreatic aquifer, northeastern part of the Algerian Sahara. Arab J Geosci 2:83–93

Said R (1962) The geology of Egypt. Elsevier, Amsterdam 377

Salamon A, Hofstetter A, Garfunkel Z, Ron H (2003) Seismotectonics of the Sinai subplate – the eastern Mediterranean region. Geophys J Int 155:149–173

Sanjuan B, Michard G, Michard A (1990) Origine des substances dissoutes dans les eaux des sources thermals et des forages de la region Asal-Ghoubbet (Republique de Djibouti). J Volcanol Geotherm Res 43:333–352

Schmitt AK, Hulen JB (2008) Burried rhyolites within the active, high-temperature Salton Sea geothermal system. J Volcanol Geotherm Res 178:708–718

Schroeder RA, Orem WH, Kharaka YK (2002) Chemical evolution of the Salton Sea, California: nutrient and selenium dynamics. Hydrobiology 473:23–45

Shalev E, Lyakhovsky V, Yechieli Y (2007) Is advective heat transport significant at the Dead Sea basin? Geofluids 7:292–300

Shalev E, Yechieli Y (2007) The effect of Dead Sea level fluctuations on the discharge of thermal springs. Isr J Earth Sci 56:19–27

Shapira A (1992) Detectability of regional seismic networks: analysis of the Israel seismic networks. Isr J Earth Sci 41:21–26

Shata AA (1982) Hydrogeology of the Great Nubian Sandstone basin, Egypt. Quart J Eng Geol Lond 15:127–133

Simon E, Mero F (1992) The salinization mechanism of Lake Kinneret. J Hydrol 138:327–343

Stanislavsky E, Gvirtzman H (1999) Basin-scale migration of continental-rift brines: paleohydrological modeling of the Dead Sea basin. Geology 27:791–794

Starinsky A (1974) Relationship between Ca-chloride brines and sedimentary rocks in Israel. Ph D Thesis Hebrew University of Jerusalem 208

Stein M (2001) The sedimentary and geochemical record of Neogene-Quaternary water bodies in the Dead Sea basin-inferences for the regional paleoclimatic history. J Paleolimnol 26:271–282

Swezey CS (1996) Structural controls on Quaternary depocentres within the Chotts Trough region of southern Tunisia. J Afr Earth Sci 22:335–347

Tarantola A, Ruegg JC, Lepine JP (1980) Geodetic evidence for rifting in Afar, 2. Vertical displacements. Earth Planet Sci Lett 48:363–370

Thiele VJ, Gramann F, Kleinsorge H (1970) Zur Geologie zwischen dem Nordrand der oestlichen Kattara-Senke und der Meetelmeer-Kueste (Aegypten, westliche Wueste). Geol Jb 88:321–354

Torfstein A, Haase-Schramm A, Waldmann M, Kolodny Y, Stein M (2009) U-series and oxygen isotope chronology of the mid-Pleistocene Lake Amora (Dead Sea basin). Geochim Cosmochim Acta 73:2603–2630

Trabelsi R, Kacem A, Zouari K, Rozanski K (2009) Quantifying regional groundwater flow between Continental Intercalaire and Djeffara aquifers in southern Tunisia using isotope methods. Environ Geol 58: 171–183

Tuttle MP, Prentice CS, Dyer-Williams K, Pena LR, Burr G (2003) Late Holocene liquefaction features in the Dominican Republic: a powerful tool for earthquake hazard assessment in the northeastern Caribbean. Seismol Soc Am Bull 93:27–46

Van Houten FB (1983) Sirte Basin, North-central Libya: Cretaceous rifting above a fixed mantle hotspot? Geology 11:115–118

Vengosh A, Starinsky A, Kolodny Y, Chivas AR (1991) Boron isotope geochemistry as a tracer for the evolution of brines and associated hot springs from the Dead Sea, Israel. Geochim Cosmochim Acta 55:1689–1695

Waldmann N, Starinsky A, Stein M (2007) Primary carbonates and Ca-chloride brines as monitors of a paleo-hydrological regime in the Dead Sea basin. Quat Sci Rev 26:2219–2228

Waldmann N, Stein M, Ariztegui D, Starinsky A (2009) Stratigraphy, depositional environments and level reconstruction of the last interglacial Lake Samra in the Dead Sea basin. Quat Res 72:1–15

Wardlaw GD, Valentine DL (2005) Evidence for salt diffusion from sediments contributing to increasing salinity in the Salton Sea, California. Hydrobiology 533:77–85

Watts JM, Swan BK, Tiffany MA, Hurlbert SH (2001) Thermal, mixing, and oxygen regimes of the Salton Sea, California, 1997–1999. Hydrobiology 466:159–176

Weinberger R, Bar-Matthews M, Levi T, Begin ZB (2007) Late-Pleistocene rise of the Sedom diapir on the backdrop of water-level fluctuations of Lake Lisa, Dead Sea basin. Quat Int 175:53–61

Winsor K, Curran HA, Greer L, Glumac B (2006) Paleoenvironmental implications of mid-Holocene serpulid tube/tufa mounds and underlying coral colonies, Enriquillo Valley, Dominican Republic. Geol Soc Am Abstr Programs 38:68

Wright EP, Benfield AC, Edmunds WM, Kitching R (1982) Hydrogeology of the Kufra and Sirte basins, eastern Libya. Quat J Eng Geol Lond 15:83–103

Wright TJ, Ebinger C, Biggs J, Ayele A, Yirgu G, Keir D, Stork A (2006) Magma-maintained rift segmentation at continental rupture in the 2005 Afar dyking episode. Nature 442:291–294

Yechieli Y (1993) The effects of water level changes in closed lakes (Dead Sea) on the surrounding groundwater and country rocks. PhD Thesis, Weizmann Institute of Science, Rehovot, Israel

Yechieli Y (2000) Fresh–saline water interface in the western Dead Sea area. Groundwater 38: 615-623

Yechieli Y (2006) Response of the groundwater system to changes in the Dead Sea level. In: Enzel Y, Agnon A, Stein M (eds) New frontiers in Dead Sea paleoenvironmental research. Geological Society of America Special Paper, Boulder, CO 401:113–126

Yechieli Y, Abelson M, Bein A, Crouvi O, Shtivelman V (2006) Sinkhole "swarms" along the Dead Sea coast: reflection of disturbance of lake and adjacent groundwater systems. Geol Soc Am Bull 118:1075–1087

Yechieli Y, Gavrieli I, Ronen D, Berkovitz B (1998) Will the Dead Sea die? Geology 26 (8):755–758

Yechieli Y, Kafri U, Goldman M, Voss CI (2001) Factors controlling the configuration of the fresh–saline water interface in the Dead Sea coastal aquifer: synthesis of TDEM surveys and numerical groundwater modeling. Hydrogeol J 9:367–377

Yechieli Y, Kafri U, Wollman S, Shalev E, Lyakhovsky V (2009a) The effect of base level changes and geological structures on the location of the groundwater divide, as exhibited in the hydrological system between the Dead Sea and the Mediterranean Sea. Geol Surv Isr Rep GSI/10/2009 22 pp

Yechieli Y, Kafri U, Wollman S, Shalev E, Lyakhovsky V (2009b) The effect of base level changes and geological structures on the location of the groundwater divide, as exhibited in the hydrological system between the Dead Sea and the Mediterranean Sea. J Hydrol 378:218–229

Yechieli Y, Ronen D, Berkovitz B, Dershovitz WS, Hadad A (1995) Aquifer characteristics derived from the interaction between water levels of a terminal lake (Dead Sea) and an adjacent aquifer. Water Resour Res 31: 893–902

Yechieli Y, Ronen D, Kaufman A (1996) The source and age of groundwater brines in the Dead Sea area, as deduced from ^{36}Cl and ^{14}C. Geochim Cosmochim Acta 60:1909–1916

Yechieli Y, Shalev E, Wollman S, Kafri U (2010) The response of the Mediterranean and Dead Sea coastal aquifers to sea level variations. Geol Surv Isr Int Rep

Yechieli Y, Starinsky A, Rosenthal E (1992) Evolution of brackish groundwater in a typical arid region: Northern Arava Rift Valley, southern Israel. Appl Geochem 7:361–374

Zak I (1967) The geology of Mount Sdom. Ph D Thesis, Hebrew University of Jerusalem (In Hebrew)

Zammouri M, Siegfried T, El-Fahem T, Kriaa S, Kinzelbach W (2007) Salinization of groundwater in the Nefzawa oases region, Tunisia: results of a regional-scale hydrogeologic approach. Hydrogeol J 15:1357–1375

Index

FSC
www.fsc.org
MIX
Papier | Fördert
gute Waldnutzung
FSC® C083411

Kolibri 360 GmbH
Ferdinand-Jühlke-Straße 7
99095 Erfurt, Deutschland
produktsicherheit@kolibri360.de